Plant Biodiversity Conservation in Ethiopia

Kflay Gebrehiwot Yaynemsa

Plant Biodiversity Conservation in Ethiopia

A Shift to Small Conservation Reserves

Springer

Kflay Gebrehiwot Yaynemsa
Samara University
Semera, Ethiopia

ISBN 978-3-031-20224-7 ISBN 978-3-031-20225-4 (eBook)
https://doi.org/10.1007/978-3-031-20225-4

This Springer imprint is published by the registered company Springer Nature Switzerland AG
The registered company address is: Gewerbestrasse 11, 6330 Cham, Switzerland

Dedicated
Tigray genocide victims
COVID-19 victims

Preface

When I was undergraduate student, we had a field excursion into different ecosystems of Ethiopia. After joining higher education institution, I used to take also students to different ecosystem types for field education. Furthermore, I had the chance to conduct my researches in different vegetation types of Ethiopia.

In those field trips and research activities, it was easily noticeable that the planet earth is significantly being transformed by anthropogenic activities; causing a devastating damage on biodiversity. As a result, most ecosystems in the regions I visited were covered by very few large forests and extremely fragmented natural vegetation. However, the focus of the trips was the intact natural forests because these were protected areas.

In the meantime, I started questioning myself "Why are we focusing on the few remnant natural forests"? "Could these few remnant forests conserve as much plant biodiversity as possible"? "What about the rare plant species found in fragmented landscape"? "Does the protected areas in Ethiopia that were established in the 1960s representative and considered the impact of climate change"?

These questions triggered me to contribute this piece of work to the Ethiopian readers in particular and global readers at large. I believe undergraduate and postgraduate students, researchers, policy, and decision-makers across several disciplines such as ecology, botany, conservation biology, biodiversity, and phytogeography would benefit the most from this book.

I believe my 15 years of teaching experience in university and my research works on several themes such as plant community ecology, montane ecology, ethnobotany, climate change, specie distribution modeling, social–ecological systems, and conservation make me the right candidate to prepare this book.

Man-induced climate change/global warming coupled with other anthropogenic factors for instance land use/land cover change are affecting biodiversity across all biomes. As a result, ecosystems are being transformed and biodiversity is threatened at an alarming rate. Most protected areas in Ethiopia were established in the 1970s without considering the aforementioned impacts. Currently, those protected areas might not be sufficient to conserve the plant biodiversity in Ethiopia. Considering small conservation reserves might benefit several threatened species which are found in smaller patches. These smaller patches could also be used for ecological connectivity for species that are shifting their niche. Therefore, this book is timely contribution.

Chapters 1 and 2 present an introduction to biodiversity, the value of biodiversity, major threats to biodiversity, especially plants and anthropogenic impact on plant biodiversity.

Chapters 3 and 4 present about plant biodiversity conservation in Ethiopia and the lost opportunities over the past decades. Some plant conservation approaches are briefly discussed in this chapter. Many species that were not included in the flora of Ethiopia are documented. Furthermore, the concept of potential natural vegetation is presented in this chapter by relating it to ecological restoration. The potentia natural vegetation and their characteristic species are also discussed.

Chapter 5 presents the Single Large or Several Small (SLOSS) conservation reserves through some case studies and forwarding some future implications. Chapter 6 introduces to pollination and dispersal across fragmented human-dominated landscapes by relating it to ecological connectivity.

Chapters 7–9 are about ecological restoration, ecological connectivity, and novel ecosystems. Here, a novel approach is presented as a way forward to ecological connectivity. Furthermore, the paradox about novel ecosystems with respect to developing countries is also discussed in these chapters.

Chapter 10 mainly presents strategic conservation planning such as systematic conservation planning (SCP) and evidence-based conservation. In addition, social–ecological systems (SESs) or coupled human-environment systems (CHES) also received the desired attention.

The last chapter presents the trends in plant ecology research in Ethiopia over the past fifty years. Plant ecology research contributes significantly to plant conservation. Due to this fact, the gaps of these plant ecological researches are clearly identified, and I implied future plant ecology research in Ethiopia that benefits plant biodiversity conservation.

Semera, Ethiopia Kflay Gebrehiwot Yaynemsa

Contents

Acronyms

AA	Afroalpine belt
ACB	*Acacia-Commiphora* woodland and bushland
AES	Abrupt changes in ecological systems
CBD	Convention on biodiversity
CTW	*Combretum-Terminalia* woodland and grassland complex
DAF	Dry evergreen Afromontane forest and grassland complex
DSS	Desert and semi-desert scrubland
EB	Ericaceous belt
EBI	Ethiopian Biodiversity Institute
EOTC	Ethiopian Orthodox Tewahido Church
FEE	Flora of Ethiopia and Eritrea
FLV	Fresh water lakes, lake shores, marshes, swamps, and flood plains vegetation
GDP	Gross domestic product
HANPP	Human appropriation of net primary productivity
IAF	Intermediate evergreen Afromontane Forest
IAS	Invasive Alien Species
IPCC	Intergovernmental Panel on Climate Change
IUCN	International Union for Conservation of Nature
m a.s.l.	Meters above sea level
MAB	Man and biosphere
MAF	Moist evergreen Afromontane forest
NBSAP	National Biodiversity Strategy and Action Plan
NFPAs	National forest priority areas
OECMs	Other effective area-based conservation measures
PNV	Potential natural vegetation
ROI	Return-on-investment
RV	Riverine vegetation
SCP	Systematic conservation planning
SDM	Species distribution models
SESs	Social–ecological systems

SLOSS	Single large or several small reserves
SLV	Salt water lakes, lake shores, salt marshes and pan vegetation
SOCS	Soil organic carbon stock
TRF	Transitional rainforest
UNEP-WCM	United Nations Environment Programme World Conservation Monitoring Centre
WDPA	World Database on Protected Area
WGG	Wooded grassland of the western Gambella Region

Chapter 1
Introduction

"Biodiversity is a concept that should be explained and not defined."
– Dan L. Perlman & Glenn Adelson, Biodiversity exploring values and priorities in conservation

Abstract Biodiversity is the variability among living organisms from all sources including, *inter alia*, terrestrial, marine, and other aquatic ecosystems and the ecological complexes of which they are part: this includes diversity within species, between species and of ecosystems. Biodiversity has multifaceted value for instance, food, economy, medicine, cultural value, aesthetic, ecological-scientific, etc. However, biodiversity is threatened at all spatial scales. Some of the major threats to biodiversity include habitat loss, habitat fragmentation, invasive alien species, over-exploitation, and fire.

Keywords Biodiversity · Biodiversity value · Climate change · Habitat loss · Habitat fragmentation · IAS

1.1 Biodiversity

The terms biodiversity and biological diversity are used interchangeably. Even though the central meaning is almost similar, researchers define biodiversity differently. However, at the last of the twenty centuries, its definition has been institutionalized via the agreement of member parties. Biodiversity means "the variability among living organisms from all sources including, inter alia, terrestrial, marine and other aquatic ecosystems and the ecological complexes of which they are part: this includes diversity within species, between species and of ecosystems" (Secretariat 1992). The important concepts in this definition are scale and complex ecological interactions.

Biodiversity has tremendous value. Nevertheless, the perspective in which the value is assessed always matters. If the perspective is human values of Biodiversity, Kellert (1993) classified as utilitarian, naturalistic, ecological-scientific, aesthetic, symbolic, humanistic, moralistic, dominionistic, and negativistic. Generally, it is possible to categorize human values of biodiversity into ecological, economic,

K. G. Yaynemsa, *Plant Biodiversity Conservation in Ethiopia*,
https://doi.org/10.1007/978-3-031-20225-4_1

cultural, and aesthetics. Human beings depend on biodiversity in every aspect of their life.

1.1.1 The Value of Biodiversity

1.1.1.1 Food, Economy, Medicine

Biodiversity is the source of food and food derivatives. These might be consumed directly or processed. Several domesticated crops and their wild relatives are the main food sources of several populations in developing countries such as Ethiopia. These domesticated crops include *Eragrostis tef* Zucc., *Triticum aestivum* L., *Sorghum bicolor* (L.) Moench, and *Coffea arabica* L. Species such as *Mimusops kummel* A. DC., *Cordia africana* Lam., *Balanites aegyptiaca* (L.) Delile., *Diospyros mespiliformis* Hochst. ex A. DC., *Ziziphus spina-christi* (L.) Desf., *Ximenia americana* L., and *Tamarindus indica* L. are reported as wild edible plants in different parts of Ethiopia.

Some plants crucially contribute significantly to the GDP of the country. In this regard, *Sesamum orientale* L. and *C. arabica* have a profound contribution to the GDP of Ethiopia. Noticeably, *Boswellia* species are the source of Myrrh globally and locally in Ethiopia. Ethiopia also exports some spice plants to support the GDP of the country.

About 80% of the drugs produced globally are extracted from plants (Fabricant and Farnsworth 2001). Several plants are globally acknowledged as important medicinal plants. These include *Artemesia* for Malaria and the Madagascar periwinkle (*Catharanthus roseus* (L.) G. Don) for cancer treatment. Without exaggeration, the Basil locally called dama kessie (*Ocimum lamiifolium* Hochst. ex. Benth.) is probably found in every household and is recognized by many Ethiopians as herbal medicine. *Ocimum urticifolium* Benth., *Jasminum abyssinicum* Hochets. Ex DC., *Withania somnifera* (L.) Dunal, *Justicia schimperiana* T. Anderson, *Allium sativum* L., and *Croton macrostachyus* Hochst. ex Delile are some of the vascular plants commonly reported as herbal medicine in Ethiopia.

1.1.1.2 Ecological-Scientific Value

Ecosystem and ecological complexes are the elements of biodiversity. Understanding the importance of biodiversity and its complicated interactions needs a scientific study. Hence, biodiversity, apart from food, economy, and medicinal values, is also an arena for scientific research. To set conservation plans and priorities, species and their interaction with other species and their environment need to be investigated. Consequently, biodiversity is focal for such kind of scientific research.

1.1.1.3 Cultural Value

The cultural value of biodiversity is indispensable. Several species or ecosystems are culturally associated with people. Some of them associate with their sovereignty by printing in their national flag (Fig. 1.1). The perfect example of the cultural value of biodiversity is the plant *Ficus sycomorus* L. (Odaa in Afan Oromo; sycamore tree) in the Oromo people, Ethiopia. This plant is symbolic in the Oromia regional state and opposition party (Oromia Liberation Front) flags. It is the symbol of peace and stability, which is believed to create a link between *Waaqaa* [God] and the people. The people believes "Waaqaa invoked spirits of dedication to big trees, rivers and big mountains and the areas where referred to as sacred" (Hinew et al. 2013). The community arrange socio-political and religious meetings under the shade of the *Odaa* tree. Furthermore, *Dobera glabra* (Forssk.) Juss. ex Poir. has also almost similar cultural value in the agro-pastoralists and pastoralists community, particularly in Afar. The oldest known tree in African, *Adansonia digitata* L., is the national tree for several countries including Angola, Madagascar, and Senegal. Many countries in the world uses plants as emblem in their national flags. Readers interested about the national trees of several countries and the associated stories might need to visit https://en.wikipedia.org/wiki/List_of_national_trees.

1.1.1.4 Aesthetic Value

Man's affiliation to nature goes back to the time immemorial. Although it might seem theistic, people envisage heaven as the most beautiful thing in their mind. When people saw a splendid environment—they say wow, it is like heaven. Although nobody has visited heaven, most people talk about it. Hence, biodiversity has magnificent aesthetic value. Most poets and artists prefer biodiversity as aesthetic value. Hence, they often visit forests and other pristine environment to make their works.

Aesthetic value of biodiversity is often judged by the human perspective. However, integrating aesthetic aspects of art, philosophy, social sciences, and cultural history could have crucial contribution to biodiversity conservation at all biodiversity scales (Tribot et al. 2018).

1.1.2 Major Threats to Biodiversity

Despite its tremendous value, biodiversity is threatened at an alarming rate. Considering the species taxonomically described and threat status assessed, it is not dubious that most species, ecosystems, and their interactions are threatened. This happens in all ecosystems albeit the magnitude difference. Similar to the magnitude difference, the underlying causes are also different. Taking into account that several species are not taxonomically described and their status unassessed, the threat could be beyond we can imagine. The major threats to biodiversity have been perceived as habitat loss,

Fig. 1.1 Examples of flags with plants as a symbol: (**a**, **c**, **f**) National, (**d**) regional, and (**e**) opposition party

habitat fragmentation, invasive alien species, over-exploitation, and fire. Recently, climate change is becoming a global phenomenon and a major threat to biodiversity.

1.1.2.1 Habitat Loss

Globally, habitat loss is the leading cause of species extinction and rarity (Sisk et al. 1994; Brummitt et al. 2015). According to Hilton-taylor (2000), habitat loss is the pervasive and overriding threat affecting 89% of all threatened birds, 83% of the threatened mammals sampled, and 91% of the threatened plants sampled (Fig. 1.2). Hence, habitat loss is an extremely deterministic phenomenon, which further determines population decline, which is one of the components of species

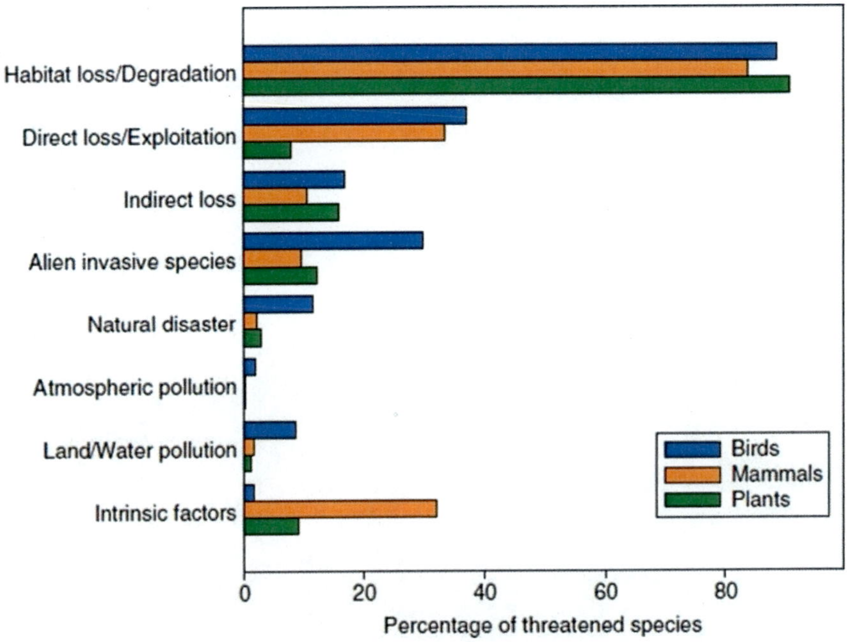

Fig. 1.2 Major threats to biodiversity (Hilton-taylor 2000)

conservation status. However, it should be noticed that habitat loss may or may not be a permanent and unidirectional problem (Lindenmayer and Fischer 2013).

The underlying causes for habitat loss could be rapid population growth, reduced incentive to conserve, and lack of enforcement, which leads to agricultural expansion, human settlement, and resource extraction. The impact of this threat is profound in developing countries such as Ethiopia. Agriculture is the leading cause of habitat loss in almost all continents.

The land use and land cover map of Ethiopia (Fig. 1.3) shows the natural ecosystem is highly transformed. A very large area of the terrestrial ecosystem is transformed into Anthropic landscapes (White 1983). Reconstruction of pre-agricultural land expansion vegetation cover of Ethiopia by Hailu et al. (2015) showed that more than one-third of today's farmlands were previously covered mainly by broad-leaved evergreen and deciduous forest. A land use/cover change study of four decades in central Ethiopia revealed that agricultural land cover is higher by 14% than the sum of forest and woodland cover (Kibret et al. 2016). A complete transformation of the woodland cover into other land cover types was also reported in other ecosystems (Garedew et al. 2009; Gebrehiwot et al. 2021).

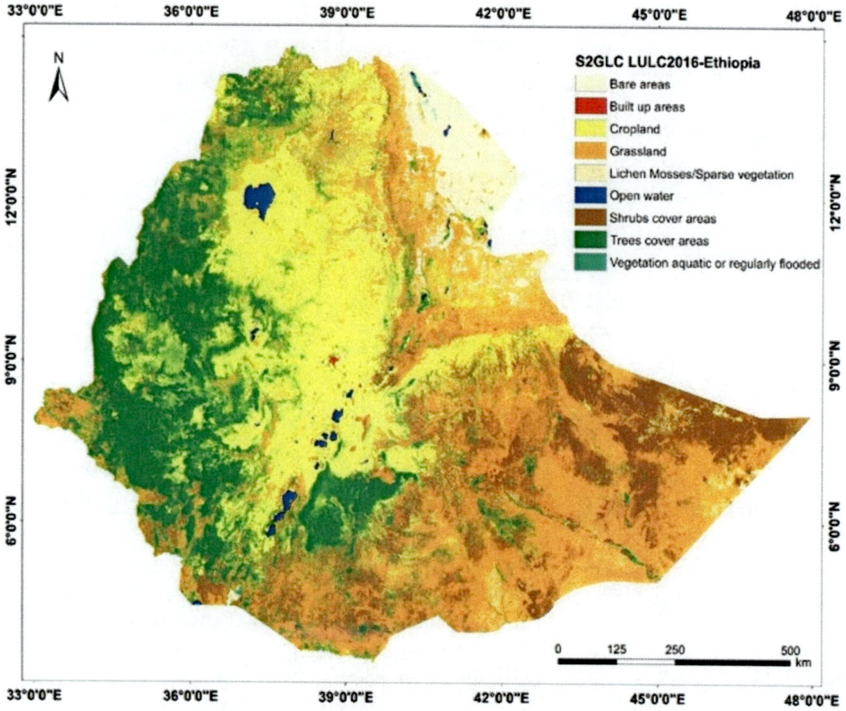

Fig. 1.3 Land use and land cover map 2016 (Hailu et al. 2018)

1.1.2.2 Habitat Fragmentation

Habitat loss and habitat fragmentation are inextricably interrelated. Habitat fragmentation could be the beginning point for habitat destruction so that it is one of the major threats to biodiversity. There is a saying in Amharic "*Wusha bekededew jib yigebal,*" which means *a tiny hole opened by a dog could be the gate for a hyena.* As clearly stated in the Amharic saying, most fragments wind up in habitat loss leading to the formation of small patches. However, Haddad et al. (2015) claimed that it is habitat loss, which leads to habitat fragmentation. Although it is obvious that habitat loss leads to fragmentation, the magnitude of loss to form fragmented habitat is not always easy to quantify. That is why it seems to contradict the previous statement.

Habitat fragmentation is the *breaking apart* of continuous habitat resulting in a reduction in the total amount of the original vegetation (i.e., habitat loss); a subdivision of the remaining vegetation into fragments, remnants, or patches (i.e., habitat fragmentation); and introduction of new forms of land use to replace vegetation that is lost (Bennett and Saunders 2010). Nevertheless, it is not possible to say habitat fragmentation if a single large area of habitat is made smaller (Fig. 1.4a) or when a whole habitat patch is removed from an area (Fig. 1.4b), because the number of

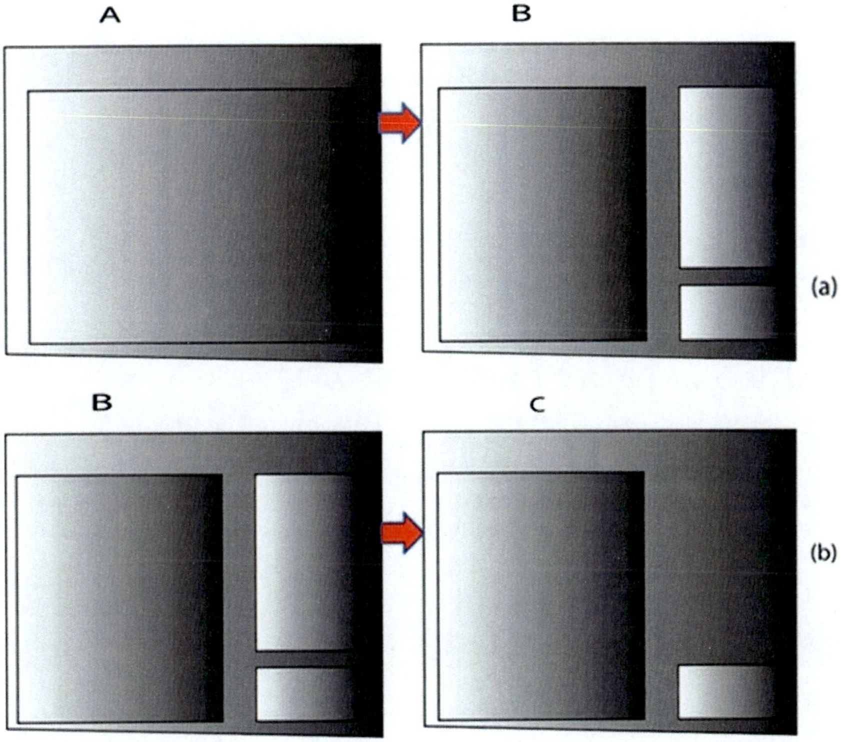

Fig. 1.4 Single large area of habitat is converted into smaller (**a**) and when a whole habitat patch is removed from an area (**b**)

patches has not increased but rather decreased (Fahrig 2019). It is due to this concept that habitat loss and habitat fragmentation are different.

Unlike the responses to habitat loss, the responses to habitat fragmentation revealed contradictory results with respect to species richness and diversity. Haddad et al. (2015) considered habitat fragmentation as one of the greatest threats to biodiversity. An experimental study of these authors reported a 20–70% decline in species richness of different taxonomic groups after fragmentation. On the contrary, Fahrig (2017) in his review titled ecological responses to habitat fragmentation per se revealed that responses to habitat fragmentation are positive. Out of the 381 studies reviewed, 290 (76%) reported a significantly positive response to habitat fragmentation, while the remaining 90 (24%) studies revealed significantly negative response to habitat fragmentation regardless whether the ecological response was a single-species or a multi-species response.

Even though the phenomenon is commonly observed in Ethiopia, habitat fragmentation studies in the country are scanty. Not only the studies are scanty but also most of them are based on theoretical justification, not experimental. For example in Abijatata-Shalla Lakes National parks, the human population residing inside the

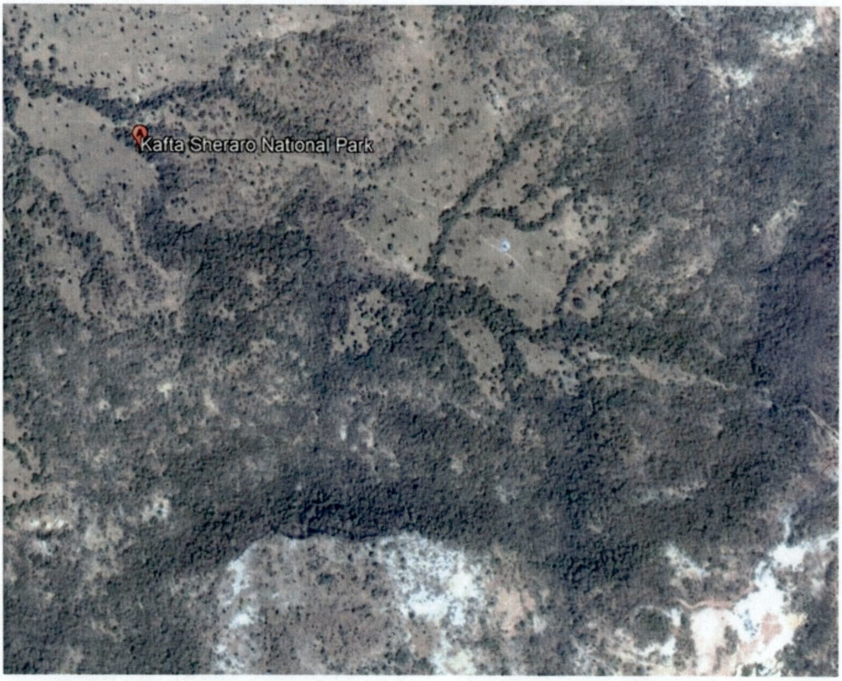

Fig. 1.5 Google Earth image of Kafta Sheraro National Park

park are nearly 44,000 (Population density = 19 People per km^2 (Teferra and Beyene 2014). Similarly, Arafaine and Asefa (2019) revealed human population size in and around the Kafta Sheraro National Park (Fig. 1.5) increased from 21,575 people in 2003 to 64,000 in 2015.

1.1.2.3 Invasive Alien Species (IAS)

Man-induced activities have transformed the biophysical nature of the earth. Often, these changes are de facto irreversible (Lenzner et al. 2019). Invasive alien species is a global phenomenon that threatens biodiversity and transforms an ecosystem. According to Simberloff (2010), invasive species arrive in a new niche, then establish a population, and spread autonomously. This is usually human-assisted. A native species might turn into invasive albeit not common. For example, a common reed (*Phragmites australis* (Cav.) Trin. ex Steud.), which was native in North America turned into invasive (Simberloff 2010).

In my teaching and research experience, I noticed that people are confused by the terms "invasive alien species" and "exotic species." I think the confusion is linked to the origin of the species—arrives in a habitat that was not occupied previously by the species. However, people should be cognizant that although both of them are exotic their performance is different. Invasive alien species aggressively invade an ecosystem and modify it. Although it is not guaranteed that exotic species would not turn into invasive ever, not all exotic species are invasive. Nevertheless, all invasive alien species are exotic. The most controversial *Eucalyptus* species is exotic. However, currently, it is not IAS in Ethiopia. On the other hand, *Prosopis juliflora* (Sw.) DC. is declared as IAS and is exotic species as well.

Generally, the myriad impact of IAS includes ecosystem modification, resource competition, predation, aggression, and its analogs, herbivory, hybridization, chain reactions, invasion meltdown, and other multiple effects (Simberloff 2010). IAS are also the main drivers of species extinction (Hilton-Taylor 2000; Bellard et al. 2016). According to Hilton-Taylor (2000), IAS is the leading cause that affected 30%, 15%, and 10% of the threatened birds, plants, and animals respectively (Fig. 1.2). In some studies, IAS was considered the leading cause for rarity following habitat destruction/loss (Wilcove et al. 1998).

Invasive alien plant species have a detrimental ecological and economic impact. Even though it is possible to quantify the economic cost, analyzing the cost of IAS on the ecosystem is challenging. Furthermore, some researchers (Nunes and Bergh 2001) criticized ecosystem cost and benefit analysis approach. These critics give sense when you try to quantify the extinction of an endemic plant by invasive plant species, what is the lost value? (Booth et al. 2010).

In Ethiopia, the issue of IAS introduction has been linked to the construction of the Ethio-Djibouti railway and road projects although a critical investigation is required. Researchers speculate that the IAS were introduced through railway and road networks [unintentionally] and some collaborative government projects [intentionally]. The invasive alien plant species *Parthenium hysterophorus* L. and *P. juliflora* are commonly mentioned in this regard. However, there is no credible source, even speculation, about the other invasive alien plant species introduction (Table 1.1).

The relatively studied invasive alien species in Ethiopia are *P. hysterophorus* and *P. juliflora*. The studies are, however, more survey and descriptive. Experiment-based studies in the country are almost none. Land cover dynamics and IAS distribution modeling, particularly, *P. juliflora* have been published (Wakie et al. 2016; Shiferaw et al. 2019). Wakie et al. (2016) estimated the current *P. juliflora* invasion is approximately 3,605 km^2 in the Afar region, while the potential habitat for future infestations is 5,024 km^2. However, more than half of the suitable habitats in that region have been already invaded. Moreover, Shiferaw et al. (2019) estimated the rate of invasion at 31,127 ha per year. Consequently, its impact has been manifested in the grassland shrinkage, native *Acacia* species population decline, and ecosystem services.

1.1.2.4 Over-Exploitation

The ever-increasing human population relies directly or indirectly on natural resources for both survival and development. As a result, biodiversity is threatened through over-exploitation. Several countries' economy, particularly the Global South

Table 1.1 Invasive plant species in Ethiopia

S. no	Species name	Family	Growth habit
1	*Acacia drepanolobium Harms ex Sjostedt*	Fabaceae	Shrub
2	*Argemone mexicana* L	Papaveraceae	Herb
3	*Calotropis procera* (Aiton) W.T.Aiton	Asclepiadaceae	Shrub
4	*Cryptostegia grandiflora* Roxb. ex R.Br	Asclepiadaceae	Subshrub
5	*Cuscuta campestris* Yunck	Cuscutaceae	Herb
6	*Eichornia crassipes* (Mart.) Solms	Pontederiaceae	Herb
7	*Lantana camara* L	Verbenaceae	Herb
8	*Nicotiana glauca* Graham	Solanaceae	Shrub
9	*Opuntia ficus-indica* (L.) Mill	Cactacea	Shrub
10	*Parkinsonia aculeata* L	Fabaceae	Shrub/tree
11	*Parthenium hysterophorus* L	Asteraceae	Herb
12	*Prosopis juliflora* (Sw.) DC	Fabaceae	Shrub/tree
13	*Psidium guajava* L	Myrtaceae	Shrub/tree
14	*Ricinus communis* L	Euphorbiaceae	Shrub/tree
15	*Senna didymobotrya* (Fresen.) H.S.Irwin & Barneby	Fabaceae	Shrub
16	*Senna occidentalis* (L.) Link	Fabaceae	Shrub
17	*Verbesina encelioides* (Cav.) Benth. & Hook.f. ex A.Gray	Asteraceae	Herb
18	*Xanthium spinosum* L	Asteraceae	Herb
19	*Xanthium strumarium* L	Asteraceae	Herb
20	*Acacia melanoxylon* R. Br.	Fabaceae	Tree
21	*Acacia saligna* Lindl.	Fabaceae	Shrub
22	*Acacia meaarnsii De Wild*	Fabaceae	Tree
23	*Typha latifolia* L.	Typhaceae	Herb

is dependent on natural resources. In Ethiopia, for example, *Coffea arabica* L. is the backbone of the economy of the country. Inappropriate investments in natural forests to support the gross domestic product (GDP) are practiced in some parts of Ethiopia. For example, the taking over of 400,000 ha of bamboo land by a foreign company in the Benishangul-Gumuz region; the taking over of vast vegetated area for the development of *Vernonia galamensis* (Cass.) Less. by a British company in the Oromia region; and several unwise investment activities in the area of the Rift Valley lakes have caused over-exploitation and pollution of lakes and wetlands in Ethiopia (Institute of Biodiversity Conservation: IBC 2009).

Boswellia papyrifera (Del.) Hochst (frankincense) is one of the over-harvested tree species found in Western Tigray, in particular, and the arid Horn in general. This is a multi-purpose tree used for medicinal, cosmetics, cultural, and religious ceremonies. The tree is tapped three times a year to collect the resin. This increases the mortality of the tree species. Consequently, *B. papyrifera* frankincense is facing extinction (Muys 2019). Bongers et al. (2019) provided accumulated evidence to add

B. papyrifera to the "vulnerable" category in the International Union for Conservation of Nature (IUCN) Red List.

Several medicinal plants are also over-exploited in developing countries such as Ethiopia. Often, medicinally important plant parts, especially, root and flower, cause the plant population to decline. A typical example is the tree *Hagenia abyssinica* (Bruce) J.F.Gmel. (African Red Wood). Traditional healers have been using the flowers of this plant for a Millennium before finalizing the reproduction stage. Hence, this caused a decline in the populations of *H. abyssinica* though the plant is not on the IUCN threatened list. Furthermore, its quality timber is also one of the reasons for its over-exploitation. Moreover, tree species such as *Prunus africana* (Hook.f.) Kalkman (Red Stinkwood or Pygeum) have been hunted for their stem bark as medicinal value. As a result, to prevent further over-exploitation, the tree was listed in the Convention on International Trade in Endangered Species of Wild Flora and Fauna (CITIES) Convention in 1995 (Duke 1997).

Indigenous tree species such as *C. africana, Podocarpus falcatus* (Thunb.) Mirb., and *Juniperus procera* Hochst. ex Endl. are extremely utilized for timber and furniture production in Ethiopia. Even though the assertion on indigenous tree species protection has been declared more than three decades, these and other species are still over-harvested. According to Hilton-Taylor (2000), over-exploitation is a threat for birds (37% of all), mammals (34% of all), and plants (8% of all) (Fig. 1.2). The issue of over-harvesting is not only the issue of developing countries but also some developed countries are also concerned. Pärtel et al. (2005) claimed that direct exploitation, particularly plant collection, is the cause for 31% of the rarity somewhere in Estonia.

1.1.2.5 Climate Change

Climate change is affecting all biomes globally. A global analysis on ecoregions claimed that climate change exacerbates the impact of habitat loss and fragmentation-climate change interactions and affected biodiversity in the ecoregions (Segan et al. 2016). The negative impact of climate change is pronounced in biomes that experience temperature and rainfall variability (Díaz et al. 2019). Plant diversity is shaped more strongly in space and time by climate than by any other factor (Harrison et al. 2020). Other taxa, ecosystems, and interactions are equally affected by climate change. Thus, biodiversity is responding differently to climate change. Plants, for example, are responding to climate change mainly in three different ways. These are species range shift, phenological shift, and phenotypic plasticity. If these responses failed to help the species survive, their fate would be extinction. These more or less also happens in other taxonomic groups in all biomes.

Recently, in their study on woody plant encroachment in Tundra and Savanna biomes, García Criado et al. (2020) revealed that woody encroachment is positively related to warming in the tundra and increased rainfall in the savanna. Hence, future conservation planning and prioritization should consider the era of man-induced climate change.

Therefore, policy- and decision-makers, conservation practitioners, researchers, and other stakeholders need to consider these climate change impacts ahead of establishing a conservation reserve and protected area networks. Species phenological shift, species range shift, and phenotypic plasticity are discussed in the following sub-sections.

Species Phenological Shift

Phenology is the seasonal pattern of activities, life histories, of an organism. It includes the critical stages of life cycles. Phenological change in response to climate change is not uncommon in several taxonomic groups and life forms of plants (Parmesan and Yohe 2003; Root et al. 2003). Variation among species in phenology is vital in avoiding competition (Cleland et al. 2007), which could help them respond better to climate change because of temperature influence (Hansen et al. 2016). However, it has to be noticed plants can also respond to other environmental factors, for example, seasonal variation in photoperiod (Edwards and Richardson 2004).

Studies revealed that the highest phenological advance occurred in early spring (Parmesan and Yohe 2003; Cleland et al. 2012). Monitoring studies tracking phenology changes due to climate change conducted by Root et al. (2003) showed an average advance of 5.1 days per decade over three decades. Similarly, IPCC (2007) reported that early spring has been advancing at a rate of 2.3–5.2 days per decade since the 1970s. Furthermore, Parmesan and Yohe (2003) reported 62% of the species in their study showed spring phenology advancement whereas 9% phenology delays in spring. The onset of autumnal phenological events was also reported somewhere else, but these shifts are less pronounced and show a more heterogeneous pattern (Walther et al. 2002), but could be noticeable as the globe is continuously warming.

Menzel and Estrella (2001); and Menzel and Fabian (1999) in Europe reported leaf color changes showed a delay of 0.3 ± 1.6 days per decade and the length of the growing season has increased by up to 3.6 days per decade over five decades. Phenological shift due to warming could increase the primary productivity of an ecosystem (Bertin 2008) since longer growing seasons are more productive than shorter growing seasons (Lieth 1974). However, plants have to control the temperature-induced resources limitations, for example, water. Defila and Clot (2001) and Menzel (2003) also reported an advancement of 1.9 days decade^{-1} and 1.6 days decade^{-1}, respectively. The estimated mean of days per decade for all species in the study, 694 species, revealed that the change in spring phenology is 5.3 ± 0.9 (Hughes and Root 2005). The interpretation of these changes is, however, not an easy task (Visser and Both 2005). Phenological shifts are influenced by altitude (Defila and Clot 2001) and expected to be more sensitive at high latitudes as well (Hughes and Root 2005). The authors claimed that the phenological shifts were predominantly recorded at higher altitudes. Considering the phenological adaptability to changing temperature, not all plants respond the same way. In species "non-responsive" to temperature, phenological behavior is regulated by a variety of different genetically controlled mechanisms (Briggs 2009).

Even though Hughes and Root (2005) claimed that phenology shift is a short-term primary response, it could influence several interactions in the community viz.

plant–pollinator interactions. Several authors reported plant–pollinator mismatches (Fabina et al. 2010). Different trophic levels may shift their phenology to different magnitudes, revealing asynchronies in timing within the ecological network (Walther 2010).

Species Range Shift

Climate is one of the determinant factors for species distribution. Species ranges shift due to climate change is determined by abiotic such as lack of sufficient environmental variables related to climate change, biotic such as competition, and biotic sustained relicts, which require a host or mutualist limited to climate change for their existence (Hampe and Jump 2011).

Since the mountains are predicted to be warmed by three times higher than the global average rate of warming recorded during the twenty century (Nogués-Bravo et al. 2007), the majority of researches on range shifts emphasizes mountains (Cannone and Pignatti 2014). This could be also the restricted movement of living organisms from mountains and pockets in extreme temperatures (Sgrò et al. 2011), which leads to extinction (Thomas et al. 2004). Geographically, the majority of the researches on plant species response to climate change has been conducted in Europe and North America mountains.

Several species are relocated from their native place without human assistance in response to climate change (Parmesan 2006). The range of species and their composition has been changing due to several factors, mainly global climate change, although the interspecific interactions within or between trophic levels could also have a crucial impact on species range shifts (Lavergne et al. 2010). However, species distribution is often influenced through species-specific physiological thresholds of temperature and precipitation tolerance (Hoffmann and Parsons 1997). The shift could be either poleward or toward elevation (Walther et al. 2002; Parmesan 2006; Lenoir et al. 2008; Chen et al. 2011).

Long-term monitoring studies revealed that changes in community plants composition might be attributed to climate change (Brooker 2010). This might lead to the formation of transient plant communities—communities that constantly change the species composition due to climate change are reported elsewhere in the world (Schippers et al. 2021). Their existence, however, depends on their ability to respond to environmental changes individually or by the emergent properties they developed. Several external factors could also cause either upward or downward shifts (Brooker 2010) of species. Hughes (2000) point out that an increase of annual temperature of 3 °C corresponds to a shift of approximately 300–400 km in latitude (in the temperate zone) or 500 m in elevation. Nonetheless, not all species respond similarly to climate change (Walther 2010).

Out of the 99 species in research of species range shift, Parmesan and Yohe (2003) reported 80% of them showed shifts in range distribution. The average poleward (latitude) and upward (elevation) shift was 6.1 ± 2.4 km and 6.1 m decade^{-1}, respectively. Chen et al. (2011), however, reported 17.6 km and 12.2 m decade^{-1} latitudinal and elevation range shifts respectively. Research in Southern California Santa Rosa Mountains by Kelly and Goulden (2008) found that the distribution of the dominant species rose by an average altitude of approximately 65 m. Except for the one species, *Agave deserti* Engelm, which shows a shift toward lower elevation by about 50 m; the remaining species showed an upward shift by about 28,142 m. Moreover, Wardle and Coleman (1992) recorded the advancement of tree line toward higher altitudes in New Zealand. Pauli (1994) also reported that the alpine plants in Europe shifted toward higher altitudes by about 1–4 m per decade due to global warming.

Lenoir et al. (2008) also studied the upward shift of an assemblage of 171 plant species along elevation gradient against climate change. They reported most of the species, one-third or 118 out of 171, shifted toward higher elevation with an average elevation of 29.4 ± 10.9 m per decade. Whereas, 53 out of 171 species shifted their range to lower elevation. Moreover, they revealed that species that shifted the most are mountainous species, which have faster life-history traits than do species showing a reduced shift (trees and shrubs), as compared with ubiquitous species (Pauli 1994; Cannone and Pignatti 2014) and increased their species composition.

Even though it is very challenging to precisely explain observed upward shifts of plant species, it could be perhaps from either upward migration through the dispersal of species from lower elevation belts and/or by resident species shifting upward (Breshears et al. 2008); range filling (without any upward shift) performed by species dispersing from existing neighbor communities within the same elevation belt (Cannone and Pignatti 2014); or because of extinctions at lower elevations; or colonization at high elevations either at the margins of or within the species' range (Wilson and Gutiérrez 2011). Krosby et al. (2015) reported the species range shift to overlap induced due to climate change. Consequently, species range shifts and the newly established communities could have an impact on the intra- and interspecific interactions that imply consequences for the functioning of ecosystems (Walther 2010). Several factors, such as habitat destruction, agricultural expansion, urbanization, may also influence species range shifts. Due to this reason, organisms may be forced to colonize a new area in which anthropogenic perturbation is minimal. Nevertheless, Grytnes et al. (2014) showed that climate warming is not the dominant driving force for species range shift and has no significant relationships.

Unlike the researches on species response to climate change, literature on plant community response to climate change is not extensive. A study on community response to climate in Vermont, USA by Pucko et al. (2011), revealed that species and community response do not follow a constant pattern rather it is complex and tends to be idiosyncratic, which indicates communities would become increasingly divergent as the magnitude of climate change increase.

Considering biological diversity at a broad scale level, biome integrity can be affected by changes in vegetation formations due to climate change (Bellard et al. 2016). Shifting dominance of species within communities could happen but also to the formation of non-analog communities, where existing species will co-occur, but in new combinations (Walther 2010). The Millennium Ecosystem Assessment forecasts there may be irreversible shifts for 5–20% of earth's terrestrial ecosystems, in particular cool conifer forests, tundra, scrubland, savannahs, and boreal forest (Lead et al. 2005). The Sahel, for example, changed from tropical forest to grassland

and then to dessert within a few thousand years (Kröpelin et al. 2008). Lapola et al. (2009) predicted large portion of the world's largest remnant forest, the Amazonian rain forest, could be converted to tropical savannah. Alpine and boreal forests are expected to expand northwards and shift their tree lines upwards at the expense of low stature tundra and alpine communities (Alo and Wang 2008). Grimm et al. (2013) reported the movement and growth of trees into adjacent Tundra.

According to (Briggs 2009) special habitat requirements, invading the already occupied territory, co-evolved mutualism (pollinators and dispersal agents), and speed and rate of dispersal play a significant role in species range shift. He also stressed that the whole plant communities do not migrate; rather the migrant plant species will establish a new plant community. This implies the present plant communities may dissociate because of plant migration in response to climate change.

Thus, species range shift and resulting community reorganizations have considerable impacts on the way species interact, and, through trophic interactions, imply consequences for the functioning of ecosystems. Hence, community reorganization will not only lead to a reshuffling of existing species; in times of global exchange of organisms but also "new" species will arrive, mix in and compose novel assemblages, and thus contribute to modified ecological networks and alter ecosystem processes (Walther 2010). This implies the need to consider climate change in establishing new conservation areas and managing the existing ones.

Phenotypic Plasticity

Besides the phenology and species range shift; plants also respond through a phenomenon called phenotypic plasticity (Anderson et al. 2012), in which range of phenotypes a single genotype can express as a function of its environment without changing its genetic constitution (Gienapp et al. 2008). This phenomenon, usually, evolves when the organisms face several abiotic and biotic conditions in their lifetime (Baythavong 2011). They can also adapt to the changes genetically through the process of evolution (Gienapp et al. 2008). It is, however, a microevolution.

Phenotypic plasticity could be either adaptive or non-adaptive (Ford-Lloyd et al. 2013). If a population can change genetically to adapt to the ever-changing climate, there could be a possibility of reducing the risk of extinction (Sgrò et al. 2011). However, accurate measurement of plasticity is very challenging (Kingsolver et al. 2012) because the genetic underpinnings of most traits are not fully known yet (Anderson et al. 2014).

Phenotypic plasticity research on three species of Patagonian steppe grasses by Couso and Fernández (2012) revealed that the species which entertain more phenotypic plasticity the better to tolerate drought. Some authors, however, argue that the phenomenon of phenotypic plasticity does not happen in many species (Merilä and Hendry 2014).

Briggs (2009) points out that under conditions of continuing climate change, it is highly likely that populations of species will reach the limits of their development adaptability and phenotypic plasticity, and organisms will be subjected to directional selection. "If a species is evolving in relation to climate at a time when major changes are occurring, then, in the simplest hypothesis, it might be expected to remain in the same geographical area without migrating" (Bradshaw and Stettler 1995). This is analogous to species shifting their range if they would become extinct if they stay in their original habitat. In his explicit review (Briggs 2009), point out there is a microevolutionary response against climate change.

Bradshaw and Stettler (1995) argued that: "*Most species, but perhaps not all, are unable to evolve, or evolve sufficiently, to cope with all aspects of the climate change. Although species may be able to evolve to some extent, they are certainly not able to evolve enough, to all the different aspects, to be able to remain in their original habitats as climate changes; they will be forced to migrate. Then, if geographical features prevent migration, they will become extinct, for example as Tsugaand Pterocaryaare in Europe.*"

Donnelly et al. (2011) argued that adaptation to climate change is heritable. Researches on garden plants in Europe, for example, Populus, revealed that the majority of the phenotypic variance in the phenology of the plants in the study could be explained by heritability (Bradshaw and Stettler 1995). However, solid evidence is still lacking whether the phenological shifts are due to phenotypic plasticity or the result of underlying genetic variability.

Overall, conservation of biodiversity in the era of man-induced climate change need to consider the major threats of biodiversity. Considering the species range shift in response to climate change, the existing conservation reserves might not be enough and establishing new nature reserves taking into account the species move would be wise and play significant role in protecting a species.

References

Alo CA, Wang G (2008) Potential future changes of the terrestrial ecosystem based on climate projections by eight general circulation models. J Geophys Res Biogeosciences 113

Anderson JT, Panetta AM, Mitchell-Olds T (2012) Evolutionary and ecological responses to anthropogenic climate change: update on anthropogenic climate change. Plant Physiol 160:1728–1740. https://doi.org/10.1104/pp.112.206219

Anderson JT, Wagner MR, Rushworth CA, et al (2014) The evolution of quantitative traits in complex environments. Heredity (Edinb) 112:4–12. https://doi.org/10.1038/hdy.2013.33

Arafaine Z, Asefa A (2019) Dynamics of land use and land cover in the Kafta-Sheraro national park, NW Ethiopia: patterns, causes and management implications. Momona Ethiop J Sci 11:239–257

Baythavong BS (2011) Linking the spatial scale of environmental variation and the evolution of phenotypic plasticity: selection favors adaptive plasticity in fine-grained environments. Am Nat 178:75–87. https://doi.org/10.1086/660281

Bellard C, Cassey P, Blackburn TM (2016) Alien species as a driver of recent extinctions. Biol Lett 12:20150623. https://doi.org/10.1098/rsbl.2015.0623

Bennett AF, Saunders DA (2010) Habitat fragmentation and landscape change. Conserv Biol All 93:1544–1550

Bertin RI (2008) Plant phenology and distribution in relation to recent climate change. J Torrey Bot Soc 135:126–146

Bongers F, Groenendijk P, Bekele T, et al (2019) Frankincense in peril. Nat Sustain 2:602–610. https://doi.org/10.1038/s41893-019-0322-2

Booth BD, Murphy SD, Swanton CJ (2010) Invasive plant ecology in natural and agricultural systems. CABI

Bradshaw HD, Stettler RF (1995) Molecular genetics of growth and development in populus. IV. Mapping QTLs with large effects on growth, form, and phenology traits in a forest tree. Genetics 139:963–973

Breshears DD, Huxman TE, Adams HD et al (2008) Vegetation synchronously leans upslope as climate warms. Proc Natl Acad Sci 105:11591–11592

Briggs D (2009) Plant microevolution and conservation in human-influenced ecosystems. Cambridge University Press

Brooker R (2010) Plant communities, plant-plant interactions and climate change. Posit plant Interact community Dyn 99–123

Brummitt NA, Bachman SP, Griffiths-lee J et al (2015) Green plants in the red: a baseline global assessment for the IUCN sampled red list index for plants. PLoS ONE 10:e0135152. https://doi.org/10.1371/journal.pone.0135152

Cannone N, Pignatti S (2014) Ecological responses of plant species and communities to climate warming: upward shift or range filling processes? Clim Change 123:201–214. https://doi.org/10.1007/s10584-014-1065-8

Chen I-C, Hill JK, Ohlemüller R, et al (2011) Rapid range shifts of species associated with high levels of climate warming. Science (80-) 333:1024–1026. https://doi.org/10.1126/science.1206432

Cleland EE, Chuine I, Menzel A et al (2007) Shifting plant phenology in response to global change. Trends Ecol Evol 22:357–365

Cleland EE, Allen JM, Crimmins TM, et al (2012) Phenological tracking enables positive species responses to climate change. Ecology 93:1765–1771. https://doi.org/10.1890/11-1912.1

Couso LL, Fernández RJ (2012) Phenotypic plasticity as an index of drought tolerance in three Patagonian steppe grasses. Ann Bot 110:849–857. https://doi.org/10.1093/aob/mcs147

Defila C, Clot B (2001) Phytophenological trends in different seasons, regions and altitudes in Switzerland. In: "Fingerprints" of climate change. Springer, pp 113–121

Díaz S, Settele J, Brondízio ES, et al (2019) Summary for policymakers of the global assessment report on biodiversity and ecosystem services of the Intergovernmental Science-Policy Platform on Biodiversity and Ecosystem Services. Intergov Sci Platf Biodivers Ecosyst Serv

Donnelly A, Caffarra A, Diskin E et al (2011) Climate warming results in phenotypic and evolutionary changes in spring events: a mini-review. Cambridge University Press, Cambridge

Duke JA (1997) The green pharmacy: New discoveries in herbal remedies for common diseases and conditions from the world's foremost authority on healing herbs. Rodale

Edwards M, Richardson AJ (2004) Impact of climate change on marine pelagic phenology and trophic mismatch. Nature 430:881–884

Fabina NS, Abbott KC, Gilman RT (2010) Sensitivity of plant–pollinator–herbivore communities to changes in phenology. Ecol Modell 221:453–458

Fabricant DS, Farnsworth NR (2001) The value of plants used in traditional medicine for drug discovery. Environ Health Perspect 109:69. https://doi.org/10.1289/EHP.01109S169

Fahrig L (2017) Ecological Responses to Habitat Fragmentation Per Se

Fahrig L (2019) Habitat fragmentation : A long and tangled tale. 33–41. https://doi.org/10.1111/geb.12839

Ford-Lloyd B, Engels JMM, Jackson M, Parry M (2013) Genetic resources and conservation challenges under the threat of climate change. Plant Genet Resour Clim Chang 16–37

García Criado M, Myers-Smith IH, Bjorkman AD, et al (2020) Woody plant encroachment intensifies under climate change across tundra and savanna biomes. Glob Ecol Biogeogr 29:925–943. https://doi.org/10.1111/geb.13072

Garedew E, Sandewall M, Soderberg U, Campbell BM (2009) Land-use and land-cover dynamics in the central rift valley of Ethiopia. Environ Manage 44:683–694. https://doi.org/10.1007/s00267-009-9355-z

Gebrehiwot K, Teferi E, Woldu Z et al (2021) Dynamics and drivers of land cover change in the Afroalpine vegetation belt: Abune Yosef mountain range, Northern Ethiopia. Environ Dev Sustain 23:10679–10701. https://doi.org/10.1007/s10668-020-01079-0

Gienapp P, Teplitsky C, Alho JS et al (2008) Climate change and evolution: disentangling environmental and genetic responses. Mol Ecol 17:167–178

Grimm NB, Chapin III FS, Bierwagen B, et al (2013) The impacts of climate change on ecosystem structure and function. Front Ecol Environ 11:474–482. https://doi.org/10.1890/120282

Grytnes J, Kapfer J, Jurasinski G et al (2014) Identifying the driving factors behind observed elevational range shifts on European mountains. Glob Ecol Biogeogr 23:876–884. https://doi.org/10.1111/geb.12170

Haddad NM, Brudvig LA, Clobert J et al (2015) Habitat fragmentation and its lasting impact on Earth's ecosystems. 1–10

Hailu B, Fekadu M, Nauss T (2018) Availability of global and national scale land cover products and their accuracy in mountainous areas of Ethiopia : a review. J Appl Remote Sens 12:041502. https://doi.org/10.1117/1.JRS.12.041502

Hampe A, Jump AS (2011) Climate relicts: past, present, future. Annu Rev Ecol Evol Syst 42:313–333. https://doi.org/10.1146/annurev-ecolsys-102710-145015

Hansen J, Sato M, Ruedy R et al (2016) Global temperature in 2015. GISS, NASA, NY. http//data giss nasa gov

Harrison S, Spasojevic MJ, Li D (2020) Climate and plant community diversity in space and time. Proc Natl Acad Sci 117:4464–4470

Hilton-Taylor C (2000) IUCN red list of threatened species. IUCN, Gland, Switzerland and Cambridge, UK

Hinew D, Odaa W, Caffee G, Galma Q (2013) Historical significances of Odaa with special reference to Walaabuu. Sci Technol Arts Res J 1:81–90. https://doi.org/10.4314/star.v1i2.98788

Hoffmann AA, Parsons PA (1997) Extreme environmental change and evolution. Cambridge University Press

Hughes L (2000) Biological consequences of global warming: is the signal already apparent? Trends Ecol Evol 15:56–61

Hughes L, Root TL (2005) Present and future phenological changes in wild plants and animals. In: Climate change and biodiversity. Yale University Press, pp 61–69

IBC (2009) Convention on biological diversity (CBD) Ethiopia's 4th Country Report. Addis Ababa, Ethiopia

IPCC (2007) Climate change 2007 the physical science basis. Cambridge University Press, USA

Kellert SR (1993) The biological basis for human values of nature. The biophilia hypothesis 42–69

Kelly AE, Goulden ML (2008) Rapid shifts in plant distribution with recent climate change. Proc Natl Acad Sci 105:11823–11826

Kibret KS, Marohn C, Cadisch G (2016) Assessment of land use and land cover change in South Central Ethiopia during four decades based on integrated analysis of multi-temporal images and geospatial vector data. Remote Sens Appl Soc Environ 3:1–19. https://doi.org/10.1016/j.rsase.2015.11.005

Kingsolver JG, Diamond SE, Siepielski AM, Carlson SM (2012) Synthetic analyses of phenotypic selection in natural populations: lessons, limitations and future directions. Evol Ecol 26:1101–1118. https://doi.org/10.1007/s10682-012-9563-5

Kröpelin S, Verschuren D, Lézine A-M et al (2008) Climate-driven ecosystem succession in the Sahara: the past 6000 years. Science (80-) 320:765–768

Krosby M, Wilsey CB, McGuire JL et al (2015) Climate-induced range overlap among closely related species. Nat Clim Chang 5:883–886. https://doi.org/10.1038/nclimate2699

Lapola DM, Oyama MD, Nobre CA (2009) Exploring the range of climate biome projections for tropical South America: the role of CO2 fertilization and seasonality. Global Biogeochem Cycles 23

Lavergne S, Mouquet N, Thuiller W, Ronce O (2010) Biodiversity and climate change: integrating evolutionary and ecological responses of species and communities. Annu Rev Ecol Evol Syst 41:321–350. https://doi.org/10.1146/annurev-ecolsys-102209-144628

Lead C, Sala OE, van Vuuren D, Zaitsev AS (2005) Biodiversity across scenarios. Ecosyst Hum Well-Being Scenar Find Scenar Work Gr 2:375

Lenoir J, Gégout J-C, Marquet PA et al (2008) A significant upward shift in plant species optimum elevation during the 20th century. Science (80-)

Lenzner B, Leclère D, Franklin O et al (2019) A framework for global twenty-first century scenarios and models of biological invasions. Bioscience 69:697–710. https://doi.org/10.1093/biosci/biz070

Lieth H (1974) Introduction to phenology and the modeling of seasonality. Phenol Seas Model 8:3–19

Lindenmayer DB, Fischer J (2013) Habitat fragmentation and landscape change: an ecological and conservation synthesis. Island Press, London, UK

Menzel A (2003) Plant phenological anomalies in Germany and their relation to air temperature and NAO. Clim Change 57:243–263

Menzel A, Fabian P (1999) Growing season extended in Europe. Nature 397:659

Menzel A, Estrella N (2001) Plant phenological changes. In: "Fingerprints" of climate change. Springer, pp 123–137

Merilä J, Hendry AP (2014) Climate change, adaptation, and phenotypic plasticity: the problem and the evidence. Evol Appl 7:1–14. https://doi.org/10.1111/eva.12137

Muys B (2019) Frankincense facing extinction. Nat Sustain 2:665–666. https://doi.org/10.1038/s41893-019-0355-6

Nogués-Bravo D, Araújo MB, Errea MP, Martinez-Rica JP (2007) Exposure of global mountain systems to climate warming during the 21st Century. Glob Environ Chang 17:420–428

Nunes PALD, van den Bergh JCJM (2001) Economic valuation of biodiversity: sense or nonsense? Ecol Econ 39:203–222

Parmesan C (2006) Ecological and evolutionary responses to recent climate change. Annu Rev Ecol Evol Syst 37:637–669

Parmesan C, Yohe G (2003) A globally coherent fingerprint of climate change impacts across natural systems. Nature 421:37–42

Pärtel M, Kalamees R, Reier Ü et al (2005) Grouping and prioritization of vascular plant species for conservation: combining natural rarity and management need. Biol Conserv 123:271–278. https://doi.org/10.1016/j.biocon.2004.11.014

Pauli GGMGH (1994) Climate effects on mountain plants. Nature 369:448

Pucko C, Beckage B, Perkins T, Keeton WS (2011) Species shifts in response to climate change: individual or shared responses? J Torrey Bot Soc 156–176

Root TL, Price JT, Hall KR et al (2003) Fingerprints of global warming on wild animals and plants. Nature 421:57–60

Schippers P, Leyequien E, Jana A (2021) Biodiversity conservation in climate change driven transient communities. Biodivers Conserv 2885–2906. https://doi.org/10.1007/s10531-021-02241-4

Secretariat CBD (1992) Convention on biological diversity. In: Convention on biological diversity

Segan DB, Murray KA, Watson JEM (2016) A global assessment of current and future biodiversity vulnerability to habitat loss-climate change interactions. Glob Ecol Conserv 5:12–21. https://doi.org/10.1016/j.gecco.2015.11.002

Sgrò CM, Lowe AJ, Hoffmann AA (2011) Building evolutionary resilience for conserving biodiversity under climate change. Evol Appl 4:326–337

Shiferaw H, Bewket W, Alamirew T et al (2019) Implications of land use/land cover dynamics and Prosopis invasion on ecosystem service values in Afar Region, Ethiopia. Sci Total Environ Sci 67:354–436. https://doi.org/10.1016/j.scitotenv.2019.04.220

Simberloff D (2010) Invasive species. Conserv Biol all 131–152

Sisk TD, Launer AE, Switky KR, Ehrlich PR (1994) Identifying extinction threats: global analyses of the distribution of biodiversity and the expansion of the human enterprise. Bioscience 44:592–602

Teferra F, Beyene F (2014) Indigenous claims and conflicts in managing the Abijata-Shalla lakes National Park, Ethiopia. Int J Biodivers Sci Ecosyst Serv Manag 10:216–227. https://doi.org/10.1080/21513732.2014.942372

Thomas CD, Cameron A, Green RE et al (2004) Extinction risk from climate change. Nature 427:145–148

Tribot AS, Deter J, Mouquet N (2018) Integrating the aesthetic value of landscapes and biological diversity. Proc R Soc B Biol Sci 285:20180971. https://doi.org/10.1098/rspb.2018.0971

Visser ME, Both C (2005) Shifts in phenology due to global climate change: the need for a yardstick. Proc R Soc B Biol Sci 272:2561–2569

Wakie TT, Hoag D, Evangelista PH et al (2016) Is control through utilization a cost effective Prosopis juliflora management strategy? J Environ Manage 168:74–86. https://doi.org/10.1016/j.jenvman.2015.11.054

Walther G (2010) Community and ecosystem responses to recent climate change. Philos Trans R Soc B 365:2019–2024. https://doi.org/10.1098/rstb.2010.0021

Walther G-R, Post E, Convey P et al (2002) Ecological responses to recent climate change. Nature 416:389–395

Wardle P, Coleman MC (1992) Evidence for rising upper limits of four native New Zealand forest trees. New Zeal J Bot 30:303–314

White F (1983) The vegetation of Africa. A descriptive memoir to accompany the UNESCO/AETFAT/UNESCO vegetation map of Africa. UNESCO, Paris, France

Wilcove DS, Rothstein D, Dubow J et al (1998) Quantifying threats to Imperiled species in the United States. Bioscience 48:607–615

Wilson RJ, Gutiérrez D (2011) Effects of climate change on the elevational limits of species ranges. Ecol consequences Clim Chang Mech Conserv Manag (EA Beever J Belant, eds) CRC Press Boca Raton, Florida 107–132

Chapter 2
Anthropogenic Impact on Plant Biodiversity

"When you get a virus, you get a fever. That's the human body raising its core temperature to kill the virus. Planet earth works the same way. Global warming is the fever, mankind is the virus. We are making our planet sick. A cull is our only hope. If we don't reduce our population ourselves, there is only one of two ways this can go. The host kills the virus, or the virus kills the host. Either way the result is the same."
Kingsman—The secret service, 2014.

Abstract Due to the man-induced changes in climate and the environment in the twenty-one century, Anthropocene epoch is proposed to be included in the geological time scale. It is suggested that the appropriate time could be 1945 onwards. In the Anthropocene epoch, anthropogenic factors or interferences influence species richness and diversity by affecting the overall resources. Even though few studies documented an increasing species richness and diversity pattern at local and regional scales, intensive anthropogenic disturbances decrease species richness and diversity. The main reasons for the increasing patterns are the introduction of non-native species, speciation (e.g., hybrid species formation with the introduced species), and global warming though it is controversial. Geographic scale should be considered to the claim that exotic species colonization increases species richness. Plant species richness and diversity are declining following the devastating forest cover change and other ecosystem transformations. Managing human disturbances by establishing nature reserves have a positive impact on improving species richness and diversity. To maintain species richness and diversity, the anthropogenic-induced disturbances that affect species loss in an ecosystem need to be managed.

Keywords Anthropocene · Anthropic communities · Diversity · Ecosystem functions · Species colonization · Species richness

2.1 What is the Anthropocene?

Geologists have generated a geological time scale based on the major life, evolutionary, and environmental events, such as glaciation, that happened over a long period. These geological time scales are classified as eras, periods, and epochs.

K. G. Yaynemsa, *Plant Biodiversity Conservation in Ethiopia*,
https://doi.org/10.1007/978-3-031-20225-4_2

Precambrian era is the oldest in the time scale and is probably the time before life appeared. However, the early Paleozoic hosted Lycopsids (including the extant club mosses and quillworts) (Boyce and Lee 2017). The eras that support life following Precambrian era are Paleozoic (Ancient), Mesozoic era (middle), and Cenozoic era (recent). These are further classified into periods and epochs. Precambrian period (Precambrian), Cambrian, Ordovician, Silurian, Devonian, Carboniferous, and Permian period (Paleozoic), Triassic, Jurassic, and Cretaceous period (Mesozoic), and the formerly called Tertiary and Quaternary now replaced by Paleogene and Neogene period (Cenozoic). Clarke et al. (2011) discuss the plant evolution and climatic events associated with the geological time scales in detail (Fig. 2.1).

The last epoch of the recent period, Neogene, is the Holocene. This epoch started about 10,000–12,000 years ago including the twenty-first Century. Due to the man-induced changes in climate and the environment in the twenty-first century, a new epoch is proposed (Crutzen and Stoermer 2000) to be included in the geological

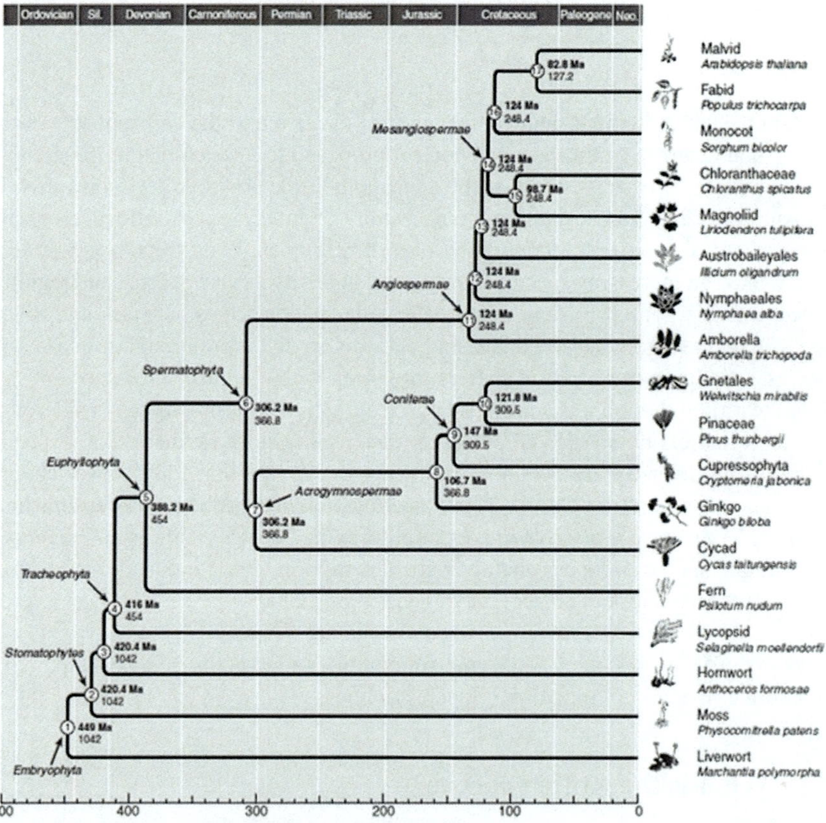

Fig. 2.1 Plant evolution across geological time scales (Clarke et al. 2011). With permission from John Wiley and Sons

time scale. This proposed epoch is called Anthropocene (anthropos "human being" and kainos "new"). It is suggested that the appropriate time could be 1945 onwards (Waters et al. 2014). Although the concept of Anthropocene, how human beings transformed the earth, is not a point of debate, the starting time for this proposed epoch raises several and complex questions. The main reason to suggest the appropriate time, 1945, is because it is on this year the atomic bomb testing began (Zalasiewicz et al. 2015) in which the isotopic by-products of the bomb testing provide a distinctive marker horizon (Ruddiman 2005).

Twelve thousand years ago, human beings performed some practices across 75% of the terrestrial environment, which in some way might have transformed the planet earth (Ellis et al. 2021). Thus, the footprints of anthropogenic impact are not the age of an old man. For example, the Ox-plow, in Ethiopia, in its current form as a dominant tool appears in rock painting dating as far back as 500 AD (McCann 2017). These counts thousands of years although it might be difficult to find stratigraphical evidence for the anthropogenic influences on biodiversity and the environment. Furthermore, agricultural land expansion in the Ethiopian highlands has been increasing since 1860 at the expense of forests, shrublands, and grasslands (Girma and Hassan 2014). According to the Intergovernmental Panel on Climate Change (IPCC 2007), about one-third of the anthropogenic CO_2 emissions since 1750 have come from land-use changes. Hence, the proposed date might be pretty convincing to geologists than ecologists and anthropologists. The Anthropocene has affected plant biodiversity in several dimensions. The following sub-sections describe the anthropogenic impacts on plant biodiversity.

2.2 Plant Biodiversity in the Anthropocene

Human beings have been influencing the planet earth for millennia. Hunter-gatherers have been using plants for different purposes for instance for food, firewood, making tools, etc. However, human beings still rely on plants for their existence. The direct and indirect exploitation of plants and environmental changes such as global warming and pollution are significantly affecting plant biodiversity. This section discusses the anthropogenic impact on plant species richness and diversity as well as exotic plant species colonization.

2.2.1 Anthropogenic Impact on Plant Species Richness and Diversity

Quantifying biodiversity is not as simple as some people might think. Because the components of biodiversity are diverse and their interaction is complex. However, species richness and diversity are some of the most used variables to determine biodiversity. Species richness and diversity are interrelated and sometimes used interchangeably by some scholars. Even though these are extremely related, they measure different attributes of biodiversity.

In simple terms, species richness is the number of species in a community or a particular area, or an ecosystem. The composition of the species determines species richness. The number of species (species richness) and their relative abundance (evenness), on the other hand, determine *diversity*. Species richness is easy to measure, unlike diversity. As a result, different authors derived several indices to measure diversity. Some of the following include the Shannon–Wiener diversity index (Shannon and Weaver 1949), Simpson's index (Simpson 1949), Pielou index (Pielou 1966), Hill numbers (Hill 1973), Brillouin index (Pielou 1969), etc. Description of these indices is not the goal of the book. Hence, readers interested in these and other indices might refer to the references listed.

In the Anthropocene epoch, anthropogenic factors or interferences influence species richness and diversity by affecting the overall resources (Storch et al. 2021). The impact could be either positive or negative. The magnitude is, however, not the same at local, regional and global scales. The species richness and diversity are declining globally. Generally, global species diversity is by far lower than that was 20,000 years ago (Williams et al. 2004). The man-induced habitat degradation, habitat destruction (loss), habitat fragmentation, and resources overexploitation caused this decline. The ever-increasing global population is causing a tremendous impact on plants. A study on the human appropriation of net primary productivity (HANPP) [measure of human intervention into the biosphere] revealed that it is about 25% (Krausmann et al. 2013). This decreases global and regional species richness and diversity (Storch et al. 2021). The HANPP reported for Ethiopia is even worst—it was about 63% in 2013—which is three times higher than the average HANPP of Africa (Grabher 2021). Grazing intensity in Ethiopia is one of the highest across the globe and has the highest HANPP (68%).

Ideally, it is possible to manage population growth. However, in practice, the efforts made to manage population and mitigate its negative effect on biodiversity in general and plant diversity in particular are not successful as expected. It is not also ethical to de-populate the already existing population across the globe. Consequently, the option left in this regard is establishing protected areas or nature reserves that minimize human interference on biological diversity. Since human beings occupy most of the terrestrial environments for different purposes, the trade-off between establishing nature reserves and human needs could not be an easy task. However, integrating the people in designing and implementing the nature reserves minimizes the burden. Furthermore, promoting sustainable utilization and working in renewable natural resources could improve the plant diversity across the globe. Failure to do so might facilitate unprecedented species extinction.

On the other hand, the species richness and diversity patterns on local and regional scales are not consistent. This is because it is driven by the magnitude of disturbances or interferences exerted at local and regional scales. Moreover, some abrupt ecosystem and land-use changes also determine the local species richness and diversity. Most studies revealed that anthropogenic disturbances decrease plant species richness and diversity at local spatial scales. For instance, studies in Bangladesh (Rahman et al. 2009), California (Williams et al. 2005), Thailand (Popradit et al. 2015), Nepal (Shrestha et al. 2013), India (Mishra et al. 2004), Brazil (Martínez-Ramosa et al. 2016; Ribeiro-Neto et al. 2016) showed that anthropogenic disturbances decrease species richness and diversity at local spatial scales. However, intermediate disturbances often increase species richness and diversity at a local scale, which is a long-established basic ecological principle. Here, anthropogenic interferences beyond optimum are considered. Even though this is the common pattern, in some occasions species richness and diversity could increase at local and regional scales because of anthropogenic disturbances. Few studies documented an increasing species richness and diversity pattern at local and regional scales (Vellend et al. 2017; Ellis 2019; Gao et al. 2020). The main reasons for the increasing patterns are the introduction of non-native species, speciation (e.g., hybrid species formation with the introduced species), and global warming though it is controversial. Nevertheless, if the introduced species are invasive, the outcome is decreasing the species richness and diversity. Furthermore, global warming could increase the net primary productivity of plants so that could favor an increase in species richness and diversity. However, if the temperature is beyond optimum plants might respond differently (Chap. 1).

In Ethiopia, studies on the anthropogenic impact on plant species richness and diversity are extremely limited. However, some research shows that the species richness and diversity are declining following the devastating forest cover change, about 40% in the 1900s to <3% in the twenty-first Century, and other natural ecosystem transformations. In one of the National Forest Priority Areas (NFPAs), Desa'a forest, species richness decreased as a result of human disturbances (Haile et al. 2021). Furthermore, some local studies also recognized that man-induced disturbances such as logging, grazing, forest clearing, etc., have a significant negative impact on species richness and diversity (Siraj et al. 2017; Gebeyehu et al. 2019; Gebrehiwot et al. 2020; Birhanu et al. 2021). Awoke and Mewded (2019) implied that managing human disturbances have a positive impact on improving species richness and diversity. That is why establishing nature reserves that restrain human disturbances is important.

2.2.2 Anthropogenic Impacts on Exotic Plant Species Colonization (Invasion)

Plants might migrate from place to place via dispersal agents. In the Anthropocene, it is accelerating faster than any time before and is expected to accelerate. Dispersal could be either short (few meters or even centimeters) or long distance (few kilometers). Apart from dispersal, human beings often introduce non-native or exotic species intentionally or unintentionally. This might cover thousands of kilometers/miles outside the biogeographic region of the species. The purpose of introducing non-native species could be for several reasons such as ecological, economic, ornamental, biotechnology, etc. In this section, the introduction of non-native species is discussed with respect to its influence on species richness and diversity.

All exotic species do not necessarily become invasive alien species (IAS), at least in a very short period although their impact could be manifested in a longer time. Exotic or non-native species and invasive species by no means have similar definitions. Hence, their impact on plant species richness and diversity is not the same. Non-native species introduced into a particular floristic region might increase the species richness and diversity of that particular region (Box 2.1). This holds true, however, if these species do not exert any significant biotic and/or related abiotic stress on other native plants. On the other hand, if the introduced species are known to have negative interactions (e.g., allelopathy) with other native plants, these species might reduce the richness and diversity of a particular area. In certain cases, this might lead to **biotic homogenization** or **monoculture**.

Box 2.1 Global species dispersal *Guizotia abyssinica* (L.f.) Cass. (Niger seed) is an annual herb, up to 2 m high. It is native to the Ethiopian highlands and is one of the most important oil crops of the country. Ethiopia is the largest producer of Niger seed oil in the world. Niger seed yields a clear high quality edible oil. In local medicine, the seeds are used against cold, leaves and fruits are used against leprosy. Roasted and pounded seeds in boiled water are drunk during fasting periods. Flowering from September to October, honeybees collect nectar and pollen and the plant is one of the most important honey sources in Ethiopia.

As depicted in the above image that was obtained from twitter on 27 September 2021, it is recorded in Ulcombe, Kent, UK. In this regard, it is possible to say the local species richness increased by one species in Ulcombe, Kent. Nevertheless, the global species richness would not increase because the species already existed in the Ethiopian and Eritrean highlands. Hence, geographic scale should be considered to the claim that exotic species colonization increases species richness.

Nevertheless, the impact of invasive alien plant species on richness and diversity is foreseeable. One of the characteristic features of invasive species is out-competing and displacing native species and modifying the ecosystem they invaded. Hence, if the resources are exhausted by the invasive species and the ecosystem is modified for the native species, the native species could become endangered, if not extinct. This might even result in native species extirpations and extinctions (Vellend et al. 2013). This implies that there is a very rare chance that the invasive alien plant species could increase the richness and diversity of a certain locality. Blackburn et al. (2020) claimed that twenty-five percent of the plant extinctions were caused by invasive alien species.

Reports on the impact of exotic/invasive species on richness and diversity are controversial. For instance, a global study reported that species invasion has increased the species richness in most regional landscapes of the world although native species loss is likely significant (Ellis et al. 2012). This is because exotic species introduction exceeds native species loss. The little flaw in this study is it considers all introduced species as "invasion," which contradicts the very definition of invasiveness. For instance, in this study (Ellis et al. 2012), crops, ornamental plants, and domesticated plants are considered as exotic invasions. It cannot be denied that some introduced species could turn into invasive, but not all. As correctly stated by Vellend et al. (2017), non-native increase species richness and diversity of a particular region or locality. Another study in California also reported that exotic invasions have increased local species diversity up to 15% (Williams et al., 2005). Similar to Ellis et al. (2012), this study also considers introduced species as an exotic invasion. However, a global synthesis on the impact of invasive species reported an increase in species richness as a result of invasion (Pysek et al. 2012).

Lower species richness in ecosystems invaded by alien species is reported worldwide. For instance, Stanek et al. (2020) reported that the invasive red oak species (*Quercus rubra* L.) decreased species richness. Similarly, a synthesis report also argue that plant invasions markedly lower species richness and diversity in a community or ecosystem where they occur (Powell et al. 2011). A study on the impact of the invasive *Prosopis juliflora* (Sw.) DC. in two east African countries, Ethiopia and Kenya, showed that an increase in the invasive species cover decreased the plant species richness and herbaceous species biomass of the ecosystem it invaded (Linders et al. 2019).

The impacts of invasive species at local, regional, and global scales does not reveal similar pattern and reports are sometimes contradicting. On small scale, relationships between exotic/invasive and native species are often negative. On the contrary, at large scales, the relationship between exotic/invasive and native species is often positive. Fridley et al. (2007) describe this phenomenon as the "invasion paradox." Since invasive species introduction is often irreversible, it needs an appropriate level of emphasis on how to manage it. This paradox must take into account in the debates of "novel ecosystem" management or restoration (Chap. 8).

2.2.3 The Anthropogenic Plant Communities

Human beings have continuously been changing the planet earth for millennials and will continue. Some researchers estimated up to 78% of the planet today is considered anthropogenic biomes (anthromes) (Ellis et al. 2010). These have created new human-derived habitats, which allow species to spread because humans actively accelerate the rates at which species immigrate into new areas through accidental and deliberate introductions (Thomas 2020). White (1983) called such landscapes "anthropic landscapes" four decades ago to indicate how human activity has changed the natural vegetation.

In light of this concept, hence, the long-established plant community concept (Clements 1916; Gleason 1926) might require a revisit beside the continuum and discrete debate. Although their perspective is not the alteration of the environment by humans, researchers have proposed different concepts of plant communities. These include Community of Individuals (CI) (Grootjans et al. 1996; Looijen and Van Andel 1999), Integrated Community (IC) (Lortie et al. 2004), and Interactive Community (IC) (Parker 2001). However, the focus of these articles is either the trophic structure or the temporal phenomenon that affects plant communities.

In the Anthropocene, nevertheless, plant communities could be by far different from the natural communities/vegetation that appeared at the beginning of the twenty centuries, which were mainly the base to develop plant community theory. This is due to habitat degradation, habitat fragmentation that is happening globally, over-exploitation, and the introduction of exotic species. The resultant communities are either metacommunities (Leibold et al. 2004), substitute communities, or novel communities (Hobbs et al. 2006, 2013). Nevertheless, long-term monitoring of plant communities in all habitat types is needed to understand the causes and consequences of Anthropocene plant biodiversity change (Vellend et al. 2017; Thomas 2020).

2.2.3.1 Metacommunity

Metacommunity is a set of local communities connected by dispersal of multiple potentially interacting species at the landscape or regional scale (Wilson 1992; Leibold et al. 2004). For obvious reasons, this is perceived as an extension of the metapopulation concept (Leibold et al. 2004). It is an important way to think about linkages between different spatial scales in ecology. Metacommunity (Fig. 2.2) per se has more or less resulted from habitat fragmentation and the concept of metacommunity theory is at the center of habitat fragmentation. This creates a fertile ground to study metacommunity and fragmentation at the local scale. However, dealing with the dynamics of metacommunity is not an easy task.

Leibold et al. (2004) proposed four metacommunity paradigms. These are neutral models, patch dynamics, mass effects, and species sorting. However, Winegardner

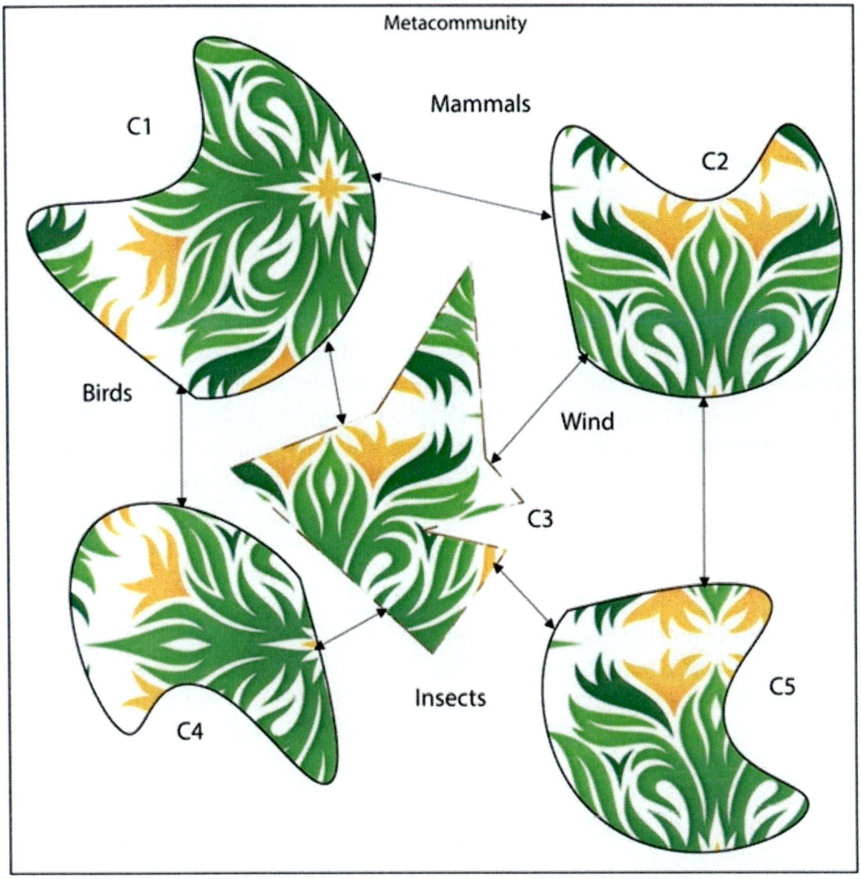

Fig. 2.2 Local communities (C1-C5) and the dispersal agents in a landscape. This demonstrates a metacommunity. The bidirectional arrows show the interactions between the local communities. Dispersal by dispersal agents (e.g., mammals, insects, and birds) is one of the mechanisms by which local communities interact

et al. (2012) argued that mass effects and patch dynamics are special cases of the species-sorting paradigm.

Neutral Model

A perspective in which all species are similar in their competitive ability, movement, and fitness.

(Hubbell 2001). *Population interactions among species consist of random walks that alter the relative frequencies of species. The dynamics of species diversity are then derived from both probabilities of species loss (extinction, emigration) and gain (immigration, speciation)* (Leibold et al. 2004).

Patch Dynamics

A perspective that assumes that patches are identical and that each patch is capable of containing populations. Patches may be occupied or unoccupied. Local species diversity is limited by dispersal. Spatial dynamics are dominated by local extinction and colonization (Leibold et al. 2004).

Mass Effect

Mechanisms for spatial dynamics in which there is a net flow of individuals created by differences in population size (density) in different patches (Wilson and Shmida 1984). A perspective that focuses on the effect of immigration and emigration on local population dynamics. In such a system, species can be rescued from local competitive exclusion in communities where they are bad competitors, by immigrating from communities where they are good competitors. This perspective emphasizes the role that spatial dynamics affect local population densities (Leibold et al. 2004).

Species Sorting

A perspective that emphasizes the resource gradients or patch types cause sufficiently strong differences in the local demography of species and the outcomes of local species interactions that patch quality and dispersal jointly affect local community composition. This perspective emphasizes spatial niche separation beyond spatial dynamics. Dispersal is important because it allows compositional changes to track changes in local environmental conditions (Leibold et al. 2004).

2.2.3.2 Substitute Community

Natural forests are extremely influenced by human land uses over centuries. Most of them are transformed into other land-use types and others are highly degraded. The plant communities associated with the original natural vegetation or potential natural vegetation (Chap. 3) are merely similar to the plant communities that occur after human interferences. Plant communities, which are composed of naturalized plant species, could never be similar to plant communities of the original vegetation. These naturalized plants, for example, replace the original vegetation that existed previously in a particular area. Thus, substitute plant communities are the communities that emerge after the original natural vegetation is lost beyond its restoration capacity to its original or historic state. It must be noted that the principal driver of substitute plant communities is prolonged human activities such as overexploitation.

Substitute plant communities are reported on several continents. For example, substitute plant communities that are resulted from human activities are reported from the Hindu Kush, Afghanistan (Breckle and Rafiqpoor 2020). Furthermore, several substitute plant communities are also reported from Bavaria, Munich (Lenz and Stary 1995). Similarly, substitute plant communities were also reported in Białowieza National Park and Laghestel di Pine‘ in Trentino (Italy) after the original plant communities were eliminated (Pedrotti 2013). Unlike permanent vegetation types, often, substitute communities are unstable. That means they are not able to withstand disturbances such as invasions.

2.2.3.3 Novel Communities

Human beings have modified the natural environment piecemeal and abruptly. Natural ecosystems such as forests, woodlands, and grasslands have been transformed into anthropic landscapes (White 1983). Consequently, the natural vegetation is either destroyed or modified to a greater extent. The introduction of exotic species changes the composition of a native plant community. Furthermore, vegetation that emerged from abandoned agricultural land could be different from native plant communities present before deforestation. The newly emerged plant communities after such situations are called novel plant communities (Lugo and Helmer 2004; Hobbs et al. 2006; Seastedt et al. 2013). This more or less resembles a mixture of introduced and native plants, which is presumably with different compositions, structures, and diversity. In northern Ethiopia, Aynekulu et al. (2012) reported a novel plant community of *Cadia purpurea-Opuntia ficus-indica* after the indigenous *Juniperus-Olea* community is extremely degraded.

2.2.4 *Anthropogenic Impact on Ecosystem Functioning*

Ecosystems provide myriad services to human beings. The notion that a reduction in biodiversity will cause a reduction in ecosystem-level processes is a well-established hypothesis (Srivastava and Vellend 2005). Often, ecosystem services are weighted in light of the benefit they provide to human beings. Hence, the trade-offs are the locus in the ecosystem service and functioning concept. Anthropogenic disturbances affect biodiversity, productivity, and ecosystem stability.

Human beings have significantly caused biodiversity loss and transformed the planet earth. This has caused a decrease in plant species richness leading to a quantitatively similar decrease in ecosystem productivity, functioning, and stability regardless of which driver caused the biodiversity loss (Tilman et al. 2006; Hautier et al. 2015). These scholars confirmed that the loss of plant diversity is linearly associated with decreased ecosystem productivity, stability, and functioning. Moreover, a study revealed that the higher plant diversity the greater ecosystem stability (Tilman et al. 2006).

Despite the number of plant species in an ecosystem, functional traits diversity also plays a crucial role in influencing ecosystem productivity, stability, and functioning (Loreau 2000). However, higher productivity does not necessarily be interpreted directly into higher levels of provision services even though its contribution to ecosystem services is crucial (Chisholm et al. 2013). The debate about the greater species diversity leads to greater ecosystem stability, productivity, and functioning is not yet settled.

Most diversity–stability relationship studies are conducted at local ecosystems. This might have contributed to the ongoing debate on the topic. Recently, however, global studies on the topic are emerging, which helps to settle the debate. Nevertheless, considering the continuously decreasing plant diversity at different spatial scales, it is possible to envisage the fate of the ecosystem's stability, productivity, and functioning. A recent study on multi-species and ecosystem functioning revealed that species richness and ecosystem functions have linearly direct relationships (Brophy et al. 2017). Moreover, an experimental study published recently revealed that as species richness [tree] increased stability of forest productivity (Schnabel et al. 2021). Hence, the loss of species could lead to the loss of ecosystem functions in either the short or the long term.

In order to either improve or maintain ecosystem functions, the species richness of an ecosystem needs to be restored, maintained, and managed. To maintain species richness, the anthropogenic-induced disturbances that affect species loss in an ecosystem need to be managed. Species loss further causes disruption of ecological networks. "If a species is lost from an ecosystem it is not just the species itself that is lost, but its interactions, and the ecological functions that result from these interactions, for example, seed dispersal" (Morris 2010). The man-induced disturbances are not simply habitat destruction, fragmentation, over-exploitation rather it extends to the current global concern, i.e., climate change. Overall, taking into account that ecosystem is dependent on the interactions in the ecosystems; maintaining diverse species as much as possible by minimizing the anthropogenic pressures on biodiversity should receive the prior attention.

References

Awoke H, Mewded B (2019) Changes in woody species composition and structure of Denkoro dry evergreen Afromontane forest over 16 years (2001–2017), South Wollo, Ethiopia. For Ecol Manage 441:71–79

Aynekulu E, Aerts R, Moonen P et al (2012) Altitudinal variation and conservation priorities of vegetation along the Great Rift Valley escarpment, northern Ethiopia. Biodivers Conserv 21:2691–2707

Birhanu L, Bekele T, Tesfaw B, Demissew S (2021) Relationships between topographic factors, soil and plant communities in a dry Afromontane forest patches of Northwestern Ethiopia. PLoS ONE 16:e0247966

Blackburn TM, Bellard C, Ricciardi A (2020) Alien versus native species as drivers of recent extinctions. Front Ecol Environ 1–5. https://doi.org/10.1002/fee.2020

Boyce CK, Lee J-E (2017) Plant evolution and climate over geological timescales. Annu Rev Earth Planet Sci 45:61–87

Breckle S-W, Rafiqpoor MD (2020) The Hindu Kush/Afghanistan. In: Plant Biogeography and Vegetation of High Mountains of Central and South-West Asia. Springer, pp 43–91

Brophy C, Dooley Á, Kirwan L et al (2017) Biodiversity and ecosystem function: making sense of numerous species interactions in multi-species communities. Ecology 98:1771–1778. https://doi.org/10.1002/ECY.1872

Chisholm RA, Muller-Landau HC, Abdul Rahman K et al (2013) Scale-dependent relationships between tree species richness and ecosystem function in forests. J Ecol 101:1214–1224. https://doi.org/10.1111/1365-2745.12132

Clarke JT, Warnock RCM, Donoghue PCJ (2011) Establishing a time-scale for plant evolution. New Phytol 192:266–301. https://doi.org/10.1111/j.1469-8137.2011.03794.x

Clements F (1916) Plant Succession: An Analysis of the Development of Vegetation. Carnegie Institution of Washington

Crutzen PJ, Stoermer EF (2000) The "Anthropocene." Glob Chang Newsl 41:17–18

Ellis EC, Goldewijk KK, Siebert S, Lightman D (2010) Anthropogenic transformation of the biomes, 1700 to 2000. Glob Ecol Biogeogr 19:589–606. https://doi.org/10.1111/j.1466-8238.2010.00540.x

Ellis EC, Gauthier N, Klein K et al (2021) People have shaped most of terrestrial nature for at least 12,000 years. Proc Natl Acad Sci U S A 118:e2023483118. https://doi.org/10.1073/pnas.2023483118

Ellis EC, Antill EC, Kreft H (2012) All Is Not Loss : Plant Biodiversity in the Anthropocene. PLoS One 7. https://doi.org/10.1371/journal.pone.0030535

Ellis EC (2019) Evolution: Biodiversity in the Anthropocene. Curr Biol 29:R829–R850. https://doi.org/10.1016/j.cub.2019.07.073

Fridley JD, Stachowicz JJ, Naeem S et al (2007) The invasion paradox: Reconciling pattern and process in species invasions. CONCEPTS Synth EMPHASIZING NEW IDEAS TO Stimul Res Ecol Ecol 88:3–17. https://doi.org/10.1890/0012-9658

Gao J-G, Liu H, Wang N, et al (2020) Plant extinction excels plant speciation in the Anthropocene. BMC Plant Biol 20:430. https://doi.org/10.1186/s12870-020-02646-3

Gebeyehu G, Soromessa T, Bekele T, Teketay D (2019) Species composition, stand structure, and regeneration status of tree species in dry Afromontane forests of Awi Zone, northwestern Ethiopia. Ecosyst Heal Sustain 5:199–215

Gebrehiwot K, Woldu Z, Fekadu M et al (2020) Classification and ordination of plant communities in Abune Yosef mountain range, northern Ethiopia. Acta Ecol Sin 40:398–411. https://doi.org/10.1016/j.chnaes.2019.12.001

Girma HM, Hassan RM (2014) Drivers of land-use change in the Southern Nations, Nationalities and People's Region of Ethiopia. African J Agric Resour Econ 9:148–164. https://doi.org/10.22004/AG.ECON.176515

Gleason H (1926) The Individualistic Concept of the Plant Association. Society 53:7–26. 10. 1007/s

Grabher H (2021) HANPP trajectories for Ethiopia reveal recent agricultural efficiency gains but high grazing intensity. Environ Dev Sustain 23:5277–5296. https://doi.org/10.1007/s10668-020-00814-x

Grootjans AP, Fresco LFM, de Leeuw CC, Schipper PC (1996) Degeneration of species-rich Calthion palustris hay meadows; some considerations on the community concept. J Veg Sci 7:185–194

Haile M, Birhane E, Rannestad MM, Adaramola MS (2021) Expansive shrubs: expansion factors and ecological impacts in Northern Ethiopia. J Nat Conserv 61:125996

Hautier Y, Tilman D, Isbell F, et al (2015) Anthropogenic environmental changes affect ecosystem stability via biodiversity. Science (80-) 348:336–340. https://doi.org/10.1126/SCIENCE.AAA1788/SUPPL_FILE/HAUTIER.SM.PDF

Hill MO (1973) Diversity and Evenness: A Unifying Notation and Its Consequences. Ecology 54:427–432. https://doi.org/10.2307/1934352

Hobbs RJ, Arico S, Aronson J et al (2006) Novel ecosystems : theoretical and management aspects of the new ecological world order. Glob Ecol Biogeogr 15:1–7. https://doi.org/10.1111/j.1466-822x.2006.00212.x

Hobbs RJ, Higgs ES, Hall CM (2013) Novel ecosystems: intervening in the new ecological world order. John Wiley & Sons, UK

Hubbell SP (2001) The unified neutral theory of biodiversity and biogeography. Princeton University Press, Princeton, NJ

IPCC (2007) Climate Change 2007 The Physical Science Basis. Cambridge University Press, USA

Krausmann F, Erb K-H, Gingrich S et al (2013) Global human appropriation of net primary production doubled in the 20th century. Proc Natl Acad Sci 110:10324–10329

Leibold MA, Holyoak M, Mouquet N et al (2004) The metacommunity concept : a framework for multi-scale community ecology. Ecol Lett 7:601–613. https://doi.org/10.1111/j.1461-0248.2004.00608.x

Lenz RJM, Stary R (1995) Landscape diversity and land use planning: a case study in Bavaria. Landsc Urban Plan 31:387–398

Linders TEW, Abebe A, Choge SK et al (2019) Direct and indirect effects of invasive species : Biodiversity loss is a major mechanism by which an invasive tree affects ecosystem functioning. J Ecol 107:2660–2672. https://doi.org/10.1111/1365-2745.13268

Looijen RC, Van Andel J (1999) Ecological communities: Conceptual problems and definitions. Perspect Plant Ecol Evol Syst 2:210–222. https://doi.org/10.1078/1433-8319-00071

Loreau M (2000) Biodiversity and ecosystem functioning: recent theoretical advances. Oikos 91:3–17. https://doi.org/10.1034/J.1600-0706.2000.910101.X

Lortie CJ, Brooker RW, Choler P et al (2004) Rethinking plant community theory. Oikos 107:433–438

Lugo AE, Helmer E (2004) Emerging forests on abandoned land: Puerto Rico's new forests. For Ecol Manage 190:145–161. https://doi.org/10.1016/j.foreco.2003.09.012

Martínez-Ramosa M, Ortiz-Rodrígueza IA, Piñerob D, et al (2016) Anthropogenic disturbances jeopardize biodiversity conservation within tropical rainforest reserves. PNAS 113:5323–5328. w ww.pnas.org/cgi/doi/https://doi.org/10.1073/pnas.1602893113

McCann JC (2017) The History of Agriculture in Ethiopia. Oxford Res. Encycl. African Hist. 1–14

Mishra BP, Tripathi OP, Tripathi RS, Pandey HN (2004) Effects of anthropogenic disturbance on plant diversity and community structure of a sacred grove in Meghalaya, northeast India. Biodivers Conserv 13:421–436

Morris RJ (2010) Anthropogenic impacts on tropical forest biodiversity: A network structure and ecosystem functioning perspective. Philos Trans R Soc B Biol Sci 365:3709–3718. https://doi.org/10.1098/RSTB.2010.0273

Parker VT (2001) Conceptual problems and scale limitations of defining ecological communities : a critique of the CI concept (Community of Individuals). Perspect Plant Ecol Evol Syst 4:80–96. 1 433-8319/01/4/02-80

Pedrotti F (2013) Plant and Vegetation Mapping. Springer, London, UK

Pielou EC (1966) The measurement of diversity in different types of biological collections. J Theor Biol 13:131–144

Pielou EC (1969) An introduction to mathematical ecology. An Introd to Math Ecol

Popradit A, Srisatit T, Kiratiprayoon S et al (2015) Anthropogenic effects on a tropical forest according to the distance from human settlements. Sci Rep 5:14689. https://doi.org/10.1038/srep14689

Powell KI, Chase JM, Knight TM (2011) A synthesis of plant invasion effects on biodiversity across spatial scales. Am J Bot 98:539–548

Pysek P, Jarosık V, Hulme PE et al (2012) A global assessment of invasive plant impacts on resident species, communities and ecosystems: the interaction of impact measures, invading species' traits and environment. Glob Chang Biol 18:1725–1737

Rahman MM, Nishat A, Vacik H (2009) Anthropogenic disturbances and plant diversity of the Madhupur Sal forests (Shorea robusta C.F. Gaertn) of Bangladesh. Int J Biodivers Sci Manag 5:162–173. 1 0.1080/17451590903236741

Ribeiro-Neto JD, Arnan X, Tabarelli M, Leal IR (2016) Chronic anthropogenic disturbance causes homogenization of plant and ant communities in the Brazilian Caatinga. Biodivers Conserv. https://doi.org/10.1007/s10531-016-1099-5

Ruddiman WF (2005) How did humans first alter global climate? Sci Am 292:46–53

Schnabel F, Liu X, Kunz M, et al (2021) Species richness stabilizes productivity via asynchrony and drought-tolerance diversity in a large-scale tree biodiversity experiment. Sci Adv 7:eabk1643. https://doi.org/10.1126/sciadv.abk1643

Seastedt TR, Hartley LM, Nippert JB (2013) Case study: ecosystem transformations along the Colorado Font Range: prairie dog interactions with multiple components of global environmental change. In: Hobbs RJ, Higgs ES, Hall M (eds) Novel Ecosystem: Intervening in the new ecological world order. John Wiley & Sons, West Sussex, UK

Shannon CE, Weaver W (1949) The mathematical theory of communication, by CE Shannon (and recent contributions to the mathematical theory of communication). University of illinois Press, W. Weaver

Shrestha KB, Måren IE, Arneberg E et al (2013) Effect of anthropogenic disturbance on plant species diversity in oak forests in Nepal, Central Himalaya. Int J Biodivers Sci Ecosyst Serv Manag 9:21–29. https://doi.org/10.1080/21513732.2012.749303

Simpson EH (1949) Measurement of diversity. Nature 163:688. https://doi.org/10.1038/163688a0

Siraj M, Zhang K, Sebsebe D, Zerihun W (2017) Floristic composition and plant community types in Maze National Park, southwest Ethiopia. Appl Ecol Environ Res 15:245–262

Srivastava DS, Vellend M (2005) Biodiversity-Ecosystem Function Research: Is It Relevant to Conservation? 101 146/annurev.ecolsys36102003152636 36:267–294. 1 0.1146/ANNUREV.ECOLSYS.36.102003.152636

Stanek M, Piechnik Ł, Stefanowicz AM (2020) Invasive red oak (Quercus rubra L.) modifies soil physicochemical properties and forest understory vegetation. For Ecol Manage 472:118253. https://doi.org/10.1016/J.FORECO.2020.118253

Storch D, Šímová I, Smyčka J et al (2021) Biodiversity dynamics in the Anthropocene: how human activities change equilibria of species richness. Ecography (cop) 44:1–19. https://doi.org/10.1111/ecog.05778

Thomas CD (2020) The development of Anthropocene biotas. Philos Trans R Soc London Ser B 375:20190113. 10.1098/r stb.2019.0113

Tilman D, Reich PB (2006) Knops JMH (2006) Biodiversity and ecosystem stability in a decade-long grassland experiment. Nat 4417093(441):629–632. https://doi.org/10.1038/nature04742

Vellend M, Baeten L, Myers-Smith IH et al (2013) Global meta-analysis reveals no net change in local-scale plant biodiversity over time. Proc Natl Acad Sci USA 110:19456–19459

Vellend M, Baeten L, Becker-Scarpitta A, et al (2017) Plant Biodiversity Change Across Scales During the Anthropocene. Annu Rev Plant Biol 68:563–86. https://doi.org/10.1146/annurev-arplant-042916- 040949

Waters CN, Zalasiewicz JA, Williams M et al (2014) A stratigraphical basis for the Anthropocene? Geol Soc 395:1–21

White F (1983) The vegetation of Africa. A descriptive memoir to accompany the UNESCO/AETFAT/UNESCO vegetation map of Africa. UNESCO, Paris, France

Williams JW, Shuman BN, Webb TI et al (2004) Late Quaternary vegetation dynamics in NorthAmerica: scaling fromtaxa to biomes. Ecol Monogr 74:309–334

Williams JW, Seabloom EW, Slayback D et al (2005) Anthropogenic impacts upon plant species richness and net primary productivity in California. Ecol Lett 8:127–137. https://doi.org/10.1111/j.1461-0248.2004.00706.x_2004

Wilson DS (1992) Complex Interactions in Metacommunities, with Implications for Biodiversity and Higher Levels of Selection. Ecology 73:1984–2000

Wilson MV, Shmida A (1984) Measuring Beta Diversity with Presence-Absence Data. J Ecol 72:1055–1064. https://doi.org/10.2307/2259551

Winegardner AK, Jones BK, Ng ISY et al (2012) The terminology of metacommunity ecology. Trends Ecol Evol 27:253–254. https://doi.org/10.1016/J.TREE.2012.01.007

Zalasiewicz J, Waters CN, Williams M et al (2015) When did the Anthropocene begin ? A mid-twentieth century boundary level is stratigraphically optimal. Quat Int J 383:196–203. https://doi.org/10.1016/j.quaint.2014.11.045

Chapter 3
The Concept of Potential Natural Vegetation (PNV)

Abstract Potential natural vegetation (PNV) explicitly refers to a definite point in time, i.e., today's PNV is likely to be different from the PNV of any point in time in the past, in which abiotic environment shaping human impact is regarded as part of the forces that have formed the environment. Lack of consensus on the definition of PNV, methods, the spatial scales used, and other variables to be considered lead to inconsistent and hardly reproducible PNV maps. Thirteen PNV are described in Ethiopia. However, due to their small coverage and overlap with other PNV types, some of them are difficult to map.

Keywords Ecosystem · Flora of Ethiopia · Vegetation mapping · Vegetation

3.1 Introduction

Vegetation has been changing both in composition and cover as a result of anthropogenic and natural disturbances. When the actual natural vegetation is transformed into another plant communities due to disturbance, understanding the historical state plays a crucial role in ecological restoration, land use, and land-use policy development. After the introduction of the concept of (PNV) by the German ecologist (Tuxen 1956) to academia in the mid-twenty century, a continuous debate has prevailed. Several scholars perceive that Tuxen (1956) defined PNV as the vegetation type that would arise in an area if all outside influences were removed [a translation from German language]. Nevertheless, a recent paper translated the German version as "PNV explicitly refers to a definite point in time, i.e., today's PNV is likely to be different from the PNV of any point in time in the past, in which abiotic environment-shaping human impact is regarded as part of the forces that have formed the environment" (Somodi et al. 2021).

PNV does not necessarily mean that the original vegetation existed in a particular area because the climatic and physical parameters may have been changed into a state that cannot be recovered. Furthermore, recent studies revealed that the concept of PNV is a cultural construct (Carrión and Fernández 2009). Chiarucci et al. (2010) criticized that the approach (sampling method, spatial scale, and projection of inferences) to characterizing PNV is unrealistic and provides scenarios with limited predictive

K. G. Yaynemsa, *Plant Biodiversity Conservation in Ethiopia*,
https://doi.org/10.1007/978-3-031-20225-4_3

power. Enriching floristic-phytosociological results by the palynological study could probably significantly contribute to the concept of PNV. This could have a crucial value to map the PNV and estimate the land for restoration and other land uses. Indeed, it is also helps to aware the public the devastating impact of land degradation and estimate land potential (Hengl et al. 2018).

Various PNV maps have been produced at different spatial scales such as national and global. However, a lack of consensus on the definition of PNV, methods, the spatial scales used, and other variables to be considered lead to inconsistent and hardly reproducible PNV maps (Zerbe 1998). Moreover, man-induced global warming could have a detrimental impact on mapping the PNV. Hence, addressing this issue with historical trends, though historical data are often unavailable, could have vital value in PNV mapping. Nevertheless, considering PNV as an operational concept, its validity should not be extrapolated either further back in time beyond the Holocene, or beyond the near future due to climate change prospects (Capelo 2018).

3.2 Potential Natural Vegetation of Ethiopia

Vegetation classification of Ethiopia was done, probably for the first time, by Pichi-Sermolli (1957) as part of northeastern Africa (Ethiopia, Eritrea, Djibouti, and Somalia) as briefly discussed in Friis and Demissew (2001). He divided the vegetation types into twenty-four. However, it was not explicit because it was based on previous geographical and botanical information. Furthermore, it was not representative at all (Friis et al. 2010).

After two decades, White (1983) classified the vegetation of Africa in general including the Ethiopian vegetation. The main criterion for the classification was mainly endemism. Out of the 80 vegetation types presented by White 13 are found in Ethiopia. The vegetation classification of White (1983) is as important as the recent vegetation classification of Ethiopia by Friis (1992) and Friis et al. (2010). However, the location of White's vegetation map lacks precision (Friis et al. 2010).

A decade later, Friis (1992) classified the vegetation of Ethiopia into six viz. (1) Lowland dry peripheral semideciduous forest, (2) transitional rain forest, (3) broad-leaved Afromontane rain forest, (4) undifferentiated Afromontane forest with either *Podocarpus falcatus* (Thunb.) C. N. Page or both *P. falcatus* and *Juniperus procera* Hochst. ex Endl., (5) dry single-dominant Afromontane forest with *J. procera*, and (6) transition between dry single-dominant Afromontane forest and east African evergreen and semi-evergreen bushland and thicket.

After Friis (1992) researchers tried to modify and classify the vegetation of Ethiopia. They classified Ethiopian vegetation/ecosystem into eight, namely (1) Afroalpine and Sub-Afroalpine, (2) dry evergreen montane forest and grassland complex, (3) moist evergreen montane forest, (4) *Acacia-Commiphora* woodland, (5) *Combretum-Terminalia* woodland, (6) lowland semi-evergreen forest, (7) desert and semi-desert scrubland, and (8) aquatic ecosystem; which was not included in the previous vegetation classification systems (Demissew et al. 1996, 2004; Woldu

1999; Friis and Demissew 2001). The former vegetation classifications were based on the actual vegetation.

Few years later, the PNV map was produced based on previous literature, field experience of the authors, and an analysis of the information of about approximately 1300 species of woody plants in the Flora of Ethiopia and Eritrea (Friis et al. 2010). This project proposed about 12 potential vegetation types with some subtypes are mapped by using GIS, remote sensing and digital elevation model (DEM) which is supplemented by some environmental variables such as altitude, topography, and rainfall patterns (Fig. 3.1; Table 3.1). Most of the vegetation types are mapped although few of them are difficult to map because of their spatial scales and overlap with other vegetation types.

The types and subtypes are (1) desert and semi-desert scrubland (DSS); (2) *Acacia-Commiphora* woodland and bushland (ACB) which has two subtypes ((2a) *Acacia-Commiphora* woodland and bushland proper (ACB), (2b) *Acacia* wooded grassland of the rift valley (ACB/RV)); (3) wooded grassland of the western Gambella Region (WGG); (4) *Combretum-Terminalia* woodland and grassland complex (CTW); (5) dry evergreen Afromontane forest and grassland complex (DAF) which have four subtypes viz. (5a) undifferentiated Afromontane forest (DAF/U), (5b) dry single-dominant Afromontane forest of the Ethiopian highlands (DAF/SD), (5c) Afromontane woodland, wooded grassland and grassland (DAF/WG) and (5d) transition between Afromontane vegetation and *Acacia-Commiphora* bushland on the eastern

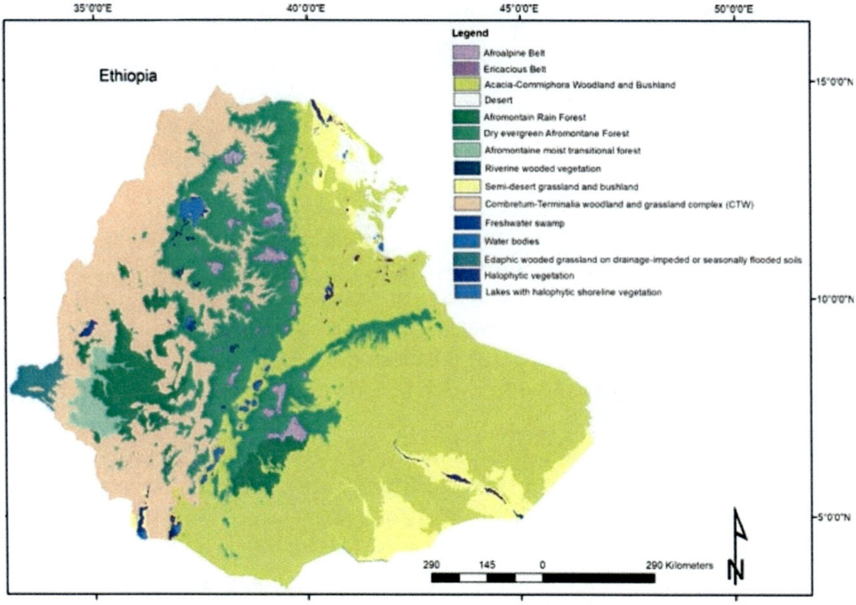

Fig. 3.1 Potential natural vegetation types of Ethiopia according to the potential natural vegetation map of East Africa (van Breugel et al. 2015)

Table 3.1 Summary of the Ethiopian vegetation types

No	Vegetation type	Elevation range	Floristic Region
1	DSS	Below 400	AF, GG, BA, SD, and HA
2	ACB	400–1900	AF, GG, and TU
	(2a) ACB proper		
	(2b) ACB/RV		
3	WGG	450–500	IL
4	CTW	400–1900	WG, TG, GD, KF, IL, GJ, and SU
5	DAF	1800–3000	
	(5a) (DAF/U)		SU, WG, WU, SD, BA and HA
	(5b) (DAF/SD)	2200–3200	TU, GD, WU and HA
	(5c) (DAF/WG)		GJ, GD, SU, AR, BA, and TU
	(5d) (DAF/TR)	1500–2000	HA, BA, SD and AR
6	MAF	1500–2600	BA, WG, SU, SD, GJ, IL, KF, and GG
	MAF/P		
	MAF/BW		
7	TRF	450–1500	WG, IL and KF
8	EB	3000–3200	GD, TU, GJ, WU, SU, GG, HA, AR, SD and BA
9	AA	Above 3200	GD, TU, GJ, WU, SU, GG, HA, AR, SD and BA, WG
10	RV	Below 1800	All
11	FLV		All
12	SLV		AF, SU and GG
13	IAF	1500–2800	GJ and GD

(AF = Afar, AR = Arsi, BA = Bale, GG = Gamo Gofa, GD = Gonder, GJ = Gojam, HA = Hararghe, IL = Illubabor, KF = Kafa, TG = Tigray, WG = Wolega, SD = Sidamo, SU = Shewa, and WU = Welo floristic regions)

escarpment (DAF/TR); (6) Moist evergreen Afromontane Forest (MAF); (7) transitional rainforest (TRF); (8) Ericaceous belt (EB); (9) Afroalpine belt (AA); (10) riverine vegetation (RV); (11) freshwater lakes, lake shores, marshes, swamps, and flood plains vegetation (FLV), and (12) salt water lakes, lake shores, salt marshes and pan vegetation (SLV). This classification system is very informative and, probably, representative.

Recently, Intermediate evergreen Afromontane Forest (IAF) is proposed as a new vegetation type of Ethiopia (Berhanu et al. 2018). This new vegetation type shares some species from dry evergreen Afromontane forest and Moist evergreen Afromontane Forest. Furthermore, their species distribution modeling revealed that Gojam (GJ) and Gondar (GD) floristic regions, with an elevation range between 1500–2800 m a.s.l., are the potential areas for this vegetation type. Nonetheless, the proposed vegetation type overlaps with the DAF and MAF. Prior to Berhanu et al.

(2018), Bekele (1994) also mentioned the transition zone between the moist southwest highlands and the relatively dry Afromontane region of the north needs due research attention. This book briefly discussed the potential natural vegetation types in Ethiopia from the perspectives of ecological restoration. This would help restoration ecologists and conservation biologists to have an insight about the historical vegetation state as well as where and what trees plant in case restoration projects through enrichment are designed.

3.2.1 Desert and Semi-Desert Scrubland (DSS)

This vegetation type is found at the lowest elevations of Ethiopia. They are dominant in the Afar floristic (AF) region, Ogaden, Gamo Gofa floristic region (GG) around Omo river basin, and the lowest extremes of Bale (BA), Sidama (SD), and Harar (HA) floristic regions at elevations 400 m below sea level. This is exclusive to the areas taken up by salt- and freshwater vegetation.

It is characterized by sparse vegetation recognized by small trees, shrubs, and herbs, which may be succulent, geophytic, or annual (Friis et al. 2010). The characteristic species of this vegetation type include—*Acacia ehrenbergiana* Hayne, *Boswellia ogadensis* Vollesen, *Commiphora erosa* Vollesen, *Gyrocarpus hababensis* Chiov., *Kissenia arabica* R. Br. ex Chiov., *Ochradenus baccatus* Delile, *Diceratella revoilii* (Franch.) Jonsell, *Cadaba barbigera* Gilg, *Ziziphus hamur* Engl., *Acacia oliveri* Vatke., *Acridocarpus glaucescens* var. *forrugineus* Engl., and *Commiphora guidottii* Chiov. ex Guid. (woody), *Euphorbia doloensis* M. G. Gilbert, *Euphorbia ogadenensis* P. R. O. Bally and S. Carter, *Euphorbia quadrispinav* S. Carter, *Aloe bertemariae* Sebsebe and Dioli, and *Aloe citrina* S. Carter and Brandham (succulents) and herbs (*Dactyloctenium aegyptium* (L.) Willd., and *Panicum turgidum* Forssk.).

3.2.2 **Acacia-Commiphora** *Woodland and Bushland (ACB)*

This vegetation type is dominantly found in the rift valley at elevations 400–900 and 1600–1900 m a.s.l. This vegetation type shares a significant number of species with the previous vegetation types, i.e., DSS. This vegetation type is the most diverse vegetation type, with the highest number of species, subspecies, and varieties not shared with other vegetation types (Friis et al. 2010).

Because of considerable differences, this complex and varied vegetation has been divided into two subtypes, one extensive subtype covering large parts of eastern and southern Ethiopia, and one restricted to part of the rift valley. These are *Acacia-Commiphora* woodland and bushland proper and *Acacia* wooded grassland of the rift valley. They are discussed independently in this book.

3.2.2.1 *Acacia-Commiphora* Woodland and Bushland Proper

This sub-vegetation type is characterized by drought-tolerant woody species that have either deciduous or small evergreen leaves. It is distributed from nearly 400 m a.s.l. to nearly 1900 m a.s.l. The characteristic species found in this sub-vegetation type include *Acacia bussei* Harms ex Y. Sjöstedt, *Acacia drepanolobium* Harms ex Y. Sjöstedt, *Acacia prasinata* Hunde (endemic), *Acacia tortilis* (Forssk.) Hayne, *Acacia zizyphispina* Chiov., *Boswellia microphylla* Chiov., *Commiphora alaticaulis* J. B. Gillett and Vollesen, *Balanites aegyptiaca* (L.) Delile, *Balanites rotundifolia* (Tiegh.) Blatt., *Boscia minimifolia* Chiov., *Cadaba ruspolii* Gilg, *Cadaba rotundifolia* Forssk., *Capparis tomentosa* Lam., *Combretum aculeatum* Vent., *Terminalia orbicularis* Engl. and Diels etc. (shrubs and trees); endemic succulents for instance *Euphorbia awashensis* M. G. Gilbert, *Aloe ellenbeckii* A. Berger, and *Aloe gilbertii* T. Reynolds ex Sebsebe and Brandham; woody climbers such as *Sarcostemma viminale* (L.) R. Br. and *Pergularia daemia* (Forssk.) Chiov.; herbaceous climbers such as *Ipomoea pogonantha* Thulin and *Ipomoea kituiensis* Vatke. Furthermore, subshrubs (*Satanocrater paradoxus* (Lindau) Lindau), Scrambling shrubs (*Asparagus aridicola* Sebsebe, *Xerophyta Schnizleinia* (Hochst.) Baker*);* herbaceous (*Chlorophytum neghellense* Cufod. (endemic), and *Drimia altissima* Hook.f. are the characteristic species of this sub-vegetation type. Trees are rarely represented in this vegetation type.

3.2.2.2 *Acacia* Wooded Grassland of the Rift Valley

This sub-vegetation type is distinguished by the Acacia species such as *Acacia albida* Delile and *A. tortilis* over the grassland. The characteristic species of this sub-type include *Acacia etbaica* Schweinf., and *Acacia seyal* Delile (trees), *Hyparrhenia, Heteropogon, Setaria, Sporobolus,* and *Panicum (herbs),* and *Aloe trichosantha* A. Berger, *A. gilbertii* (endemic), and *Euphorbia nigrispinioides* M. Gilbert (succulents).

3.2.3 Wooded Grassland of the Western Gambela Region (WGG)

This vegetation type entertains annual fire and flooding. Hence, most of the plant species in this vegetation type are fire-tolerant. Unlike the other vegetation types, most of the grasses that are found here grow very tall, sometimes above two meters. Even though the species recorded from this vegetation type are very small (seven), most of the species (five) are shared with the adjacent vegetation type, i.e., *Combretum-Terminalia woodland and wooded grassland.* It is also estimated to have very low plant diversity.

A. seyal, A. nilotica, Hyphaene thebaica (L.) Mart., *Borassus aethiopum* Mart.are abundantly found in this vegetation type. Furthermore, grasses and herbs such as *Echinochloa, Setaria, Hyparrhenia, Cymbopogon, Sorghum, Oryza longistaminata* A. Chev. and Roehr., *Caperonia serrata* (Turcz.) C. Presl, and *Thalia geniculate* L. are commonly found in this vegetation type.

3.2.4 Combretum-Terminalia *Woodland and Wooded Grassland (CTW)*

Small to medium-size trees with relatively large deciduous leaves, particularly species of the genera *Combretum* and *Terminalia* are common in these vegetation types together with other species such as *Lonchocarpus laxiflorus* Guill. and Perr., and *Piliostigma thonningii* (Schumach.) Milne-Redh. Like the WGG vegetation type, this vegetation type also experiences burning during the dry seasons. Consequently, some plant species are adapted to frequent burns.

This vegetation type hosts moderately high plant diversity. It shares a significantly high number of species with the DAF (74), and ACB and CTW (50). Except with the aquatic vegetation, it also shares several species with the other vegetation types.

The characteristic species of this vegetation types include trees and shrubs such as *Cussonia arborea, Boswellia papyrifera* (Caill. ex Delile) Hochst., *Anogeissus leiocarpa* (DC.) Guill. and Perr., *Combretum hartmannianum* Schweinf. (near-endemic), *Combretum mole* R. Br. ex G. Don, *Combretum rochetianum* A. Rich. ex (near-endemic), *Terminalia laxiflora* Engl., *Pterocarpus lucens* Lepr. ex Guill. and Perr., *Dalbergia melanoxylon* Guill. and Perr., *P. thonningii, B. aegyptiaca, Stereospermum kunthianum* Cham., *Lannea schweinfurthii* (Engl.) Engl., *Ozoroa insignis* Delile, *Acacia hockii* De Wild., *Grewia mollis* Juss., *Adansonia digitata* L. The solid-stemmed lowland bamboo *Oxytenanthera abyssinica* (A. Rich.) Munro is prominent in river valleys of this vegetation type. The ground cover is a tall stratum of perennial grasses, including *Panicum maximum* Jacq. and species of *Cymbopogon, Hyparrhenia, Echinochloa, Sorghum,* and *Pennisetum.* Others such as *Chlorophytum herrmannii* Nordal and Sebsebe (endemic); *Crinum bambusetum* Nordal and Sebsebe (endemic), *Pancratium centrale* (A. Chev.) Traub, and *Asparagus flagellaris* (Kunth) Baker are also commonly found in this vegetation type.

This vegetation type is widely distributed across the western escarpment of Ethiopia, which extends from Eritrea to Sudan borders. There is a slight elevation range overlap with the ACB vegetation type. This vegetation type is distributed between 500 and 1900 m a.sl.

3.2.5 Dry Evergreen Afromontane Forest and Grassland Complex (DAF)

This is the most transformed vegetation type in the Ethiopian highlands. It is a very complex system of successions involving extensive grasslands (Friis et al. 2010). Agricultural activities were practiced in this vegetation type for thousands of years (Woldu 1999). It is distributed in almost all floristic regions at an elevation of 1800–3000 m a.s.l.

Even though they are shared with the adjacent vegetation types, it has the second-highest plant diversity with nearly 500 species. *J. procera, Croton macrostachyus* Hochst. ex Delile, *P. falcatus, Olea europaea* subsp. *Cuspidata* (Wall. and G. Don) Cif. etc. are the characteristic species of this vegetation type.

3.2.5.1 Undifferentiated Afromontane Forest

The undifferentiated Afromontane forests are either *Juniperus-Podocarpus* forests or tend toward single-dominant *Podocarpus* or *Juniperus* forests, both with a dement of broad-leaved species (Friis et al. 2010). There are three sub-types under this vegetation type namely dry single-dominant Afromontane forest of the Ethiopian highlands (DAF/SD); Afromontane woodland, wooded grassland, and grassland (DAF/WG); and transition between Afromontane vegetation and *Acacia-Commiphora* bushland on the eastern escarpment (DAF/TR).

3.2.6 Moist Evergreen Afromontane Forest (MAF)

This vegetation type is known for its one or more closed strata of evergreen trees that can reach up to 40 m. However, due to anthropogenic disturbances, for instance, removal of the tree large tree species, the lower strata may be the remnant. It is widely distributed in southwestern Ethiopia at elevations ranging from about 1500 to 3000 m a.s.l. This vegetation type is further classified into two sub-types namely primary or mature secondary Moist evergreen Afromontane Forest (MAF/P) and edges of Moist evergreen Afromontane Forest, bushland, woodland, and wooded grassland (MAF/BW).

Most of the species found in MAF are shared with the adjacent vegetation, especially with the DAF and riverine vegetation. The characteristic species of MAF include *Pouteria adolfi-friederici* (Engl.) A. Meeuse, *Albizia gummifera* (J. F. Gmel.) C. A. Sm., *Cassipourea malosana* (Baker) Alston, *Celtis africana* Burm.f., *Sapium ellipticum* (Hochst.) Pax, *Macaranga capensis* (Baill.) Sim, *C. macrostachyus, Euphorbia ampliphylla* Pax, *Ekebergia capensis* Sparrm., *Olea welwitschia* (Knobl.) Gilg and G. Schellenb., *Polyscias fulva* (Hiern) Harms, *Schefflera abyssinica, Prunus africana* (Hook.f.) Kalkman, and *Syzygium guineense* (Willd.) DC.

3.2.7 Transitional Rainforest (TR)

This vegetation type is distributed at an elevation range between 450 and 1500 m a.s.l., and is one of the controversial classifications due to its poorly delineated boundary. It is mainly found in the southwestern Ethiopia escarpment, Ilubabor, Wollega, and Kefa floristic regions. The characteristic species of this vegetation type include *Manilkara butugi* Chiov., *Pouteria altissima* (A. Chev.) Baehni, *Milicia excelsa* (Welw.) C. C. Berg, *Morus mesozygia* Stapf, *Trilepisium madagascariense* DC., *Alstonia boonei* De Wild., *Croton sylvaticus* Schltdl., *Diospyros abyssinica* (Hiern) F. White, *Albizia schimperiana* Oliv., *Urera trinervis* (Hochst.) Friis and Immelman, *Alchornea laxiflora* (Benth.) Pax and K. Hoffm., *Argomuellera macrophylla* Pax, etc.

3.2.8 Ericaceous Belt (EB)

Despite the differences in the lower and upper limits, this vegetation type is dominantly found in the East African Mountains. This was noticed seven decades ago by Hedberg (1951). In Ethiopia, the elevation range of this vegetation type is dominantly between 3000 and 3200 m a.s.l. However, Ericaceous forest patches might be encountered at elevations either below or above the specified range. Furthermore, different terminologies have been used to represent this vegetation type. For instance, “Ericaceous belt” (Hedberg 1964; Friis et al. 2010), “ Ericaceous bushland” (White 1983) or “Ericaceous scrub” (Egziabher 1988; Demissew et al. 1999). In northern Ethiopia, this vegetation type is found in a very small remnant natural vegetation within a very steep slope. Hence, it is a disturbed and fragmented landscape which makes using the term “Ericaceous belt” difficult. Thus, some authors used the term Ericaceous forest (Gebrehiwot et al. 2019; Gebrehiwot 2020).

The characteristic species of this vegetation type includes *Erica arborea L., Hypericum revolutum* Vahl, *Myrsine melanophloeos* (L.) R. Br. ex Sweet, *Alchemilla haumannii* Rothm., *Geranium arabicum* Forssk., *Anthemis tigreensis* J. Gay ex A. Rich., *Erigeron alpinus* L., *Haplocarpha rueppellii* K. Lewin, *Helichrysum citrispinum* Delile, *Helichrysum splendidum* Less., *Helichrysum gofense, Helichrysum formosissimum* Sch.Bip., *Senecio schultzii* Hochst. ex A. Rich., *Romulea fischeri* Pax, *Satureja biflora* (Buch.-Ham. ex D. Don) Briq., *Thymus schimperi* Ronniger, *Trifolium acaule* Steud. ex A.Rich., *Trifolium burchellianum* Ser., *Aloe steudneri* Schweinf., *Kniphofia foliosa* Hochst. and *Primula verticillata* Forssk. (Friis et al. 2010). Nevertheless, considerable characteristic species of the EB were found to be a characteristic species of the Afroalpine vegetation at an elevation above 3700 m a.s.l. in some mountains such as Abune Yosef. These include *A. tigreensis, H. rueppellii, H. citrispinum, Senecio schultzii, K. foliosa* and *P. verticillata*. Several species are shared with the neighboring vegetation types; dry evergreen Afromontane forests and the Afroalpine belt.

This vegetation type is found in most of the Ethiopian floristic regions. For instance, it is found in Gonder, Tigray, Gojam, Welo, Shewa, Gamo Gofa, Harar, Arsi, Sidamo, and Bale floristic regions.

3.2.9 Afroalpine Belt (AA)

The area of this vegetation type is relatively small, which is limited to elevations above 3200 m. It is distributed across GD, TU, GJ, WU, SU, GG, HA, AR, SD, and BA, WG floristic regions although the large landmass is found in Bale and GD floristic regions.

The Afroalpine belt is characterized by its large numbers of geographically vicarious taxa and high endemic plant species (Hedberg 1969). The Afroalpine belt hosts plant species with life forms often different from the other vegetation types (Hedberg 1964). These life forms are giant rosette plants, tussock grasses (and sedges), acaulescent rosette plants, cushion plants, and sclerophyllous shrubs (and dwarf shrubs). Although Friis et al. (2010) reported that the lower belt is dominated by small trees and shrubs, the Afroalpine vegetation in the present study which is dominantly above 3700 m a.s.l. is above the tree line. However, some tree species such as *J. procera* and *Hagenia abyssinica* (Bruce) J. F. Gmel. were observed around Abune Yosef monastery and homesteads at an elevation about 3700 m a.s.l. (Gebrehiwot 2020). The characteristic species of this vegetation type are mostly herbs such as the giant *Lobelia rhynchopetalum* Hemsl., *Carex monostachya* A. Rich., and *Festuca (F. gilbertiana* E. B. Alexeev ex S. M. Phillips, *F. macrophylla* Hochst. ex A. Rich.), *Poa (P. hedbergii* S. M. Phillips, *P. chokensis* S. M. Phillips, *P. pumilio* Hochst., *P. simensis* Hochst. ex A. Rich.), and *Agrostis (A. gracilifolia* subsp. *parviflora* S. M. Phillips*)*.

3.2.10 Riverine Vegetation (RV)

Ethiopia has many inland and cross boundary rivers (e.g., Nile, Mereb, Wabi Shebele, Tekeze, etc.). Riverine vegetation is found in all floristic regions. These rivers and their tributaries are accompanied by riverine forests at an elevation often below 1800 m. This vegetation type has tall trees with evergreen leaves. Most riverine and riparian habitats are probably too narrow and striplike to map (Friis et al. 2010). A significant number of species are shared with the DAF and MAF.

The characteristic species of this vegetation type include *Trema orientalis* (L.) Bl., *Oncoba spinosa* Forssk., *Diospyros mespiliformis* Hochst. ex A. DC., *Acacia polyacantha* subsp. *campylacantha, Celtis ajricana* Burm.f., *Ficus sycomorus* L., *Mimusops kummel* Bruce ex A. DC., *Mimusops laurifolia* (Forssk.) Friis, *Breonadia salicina* (Vahl) Hepper and J. R. I. Wood, *Lepisanthes senegalensis* (Poir.) Leenh., *Salix subserrata* Willd., *Trichilia emetica* Vahl, *D. mespiliformis, S. guineense* subsp. *guineense*, T*amarindus indica* L., *Hyphaene thebaica* (L.) Mart. and *Phoenix reclinate* Jacq. and *Zanthoxylum gilletii* (De Wild.) P. G. Waterman.

3.2.11 Freshwater Lakes (FLV)

The distinction between fresh and saltwater is the amount dissolved in the water. Freshwater lakes vegetation are also commonly found in almost all floristic regions. This vegetation type is further classified into Freshwater Lake vegetation (open water) and Freshwater marshes and swamps, floodplains, and lakeshore vegetation.

In comparison to the previously discussed vegetation types, this vegetation type hosts lower species richness. The characteristic species of this vegetation type include *Lemna aequinoctalis* Welw., L. *gibba* L., *L. minor* L., *Wolffia arrhiza* (L.) Horkel ex Wimm., *Pistia stratiotes* L., *Eichornia crassipes* (Mart.) Solms, *Aphanothece microspora, Chroococus disperses, Closterium spp., Meliorosa granulata, Cyperns digitatus, C. denudatus, C. dichroostachys, C. elegantulus, C. latifolius, Ascolepis capensis, Juncus dregeanus, Floscopa glomerata, Eriocaulon* spp., *Syngonanthus wahlbergii, Xyris capensis, Persicaria decipiens, Ludwigia abyssinica, Chenopodium album, Ranunculus multibus, Plectranthus punctatus, Leersia hexandra, Nymphaea lotus, Phoenix reclinata, Aeschynomene cristata* var. *pubescens, Aeschynomene elaphroxylum, Aeschynomene pfundii* and *Aeschynomene schimperi.*

3.2.12 Salt Lakes, Salt-Lake Shores, Marsh and Pan Vegetation (SLV)

The salt lakes vegetation type is dominantly found in the rift valley lakes in AF, SU, and GG floristic regions. This vegetation type consists of two sub-types. These are salt lake vegetation (open-water), and saltpans, saline/brackish and intermittent wetlands, and salt lake shore vegetation.

Suaeda monoica Forssk. ex J. F. Gmel., *Anabaenopsis* spp., *Anomoeoneis sphaerocarpa, Oscillatoria, Atriplex* spp. and *Salicomia* spp. spp., *Spirulina platensisis,* etc. are some of the characteristic species reported in from this vegetation type.

3.2.13 Intermediate Evergreen Afromontane Forest (IAF)

This newly proposed vegetation type is was published eight years after the publication of the Atlas of the Potential Vegetation of Ethiopia (Berhanu et al. 2018). It is distributed in a narrow landmass around the Gojam and Gonder floristic regions at an elevation overlapping with the DAF and MAF, 1500–2800 m. Most of the characteristic species of this vegetation types for instance *A. schimperiana, A. gummifera, C. africana, C. tomentosa, Dalbergia lacteal* Vatke, *Clematis longicauda* Steud. ex A. Rich. and *C. simensis* Fresen. and *P. africana,* are shared with the DAF and MAF. One of the main reasons proposed as a new vegetation type is the absence of DAF and MAF characteristic species such as *O. europaea* subsp. *cuspidata* and *J. procera* in this vegetation type, unlike the MAF and DAF.

References

Bekele T (1994) Phytosociology and ecology of a humid Afromontane forest on the Central Plateau of Ethiopia. J Veg Sci 5:87–98. https://doi.org/10.2307/3235642

Berhanu A, Demissew S, Woldu Z et al (2018) Intermediate evergreen Afromontane forest (IAF) in northwestern Ethiopia: observations, description and modelling its potential distribution. Phytocoenologia 48:351–367. https://doi.org/10.1127/phyto/2018/0207

Capelo J (2018) Evidence of a unique association between single forest vegetation-types and seral sequences: praise for the concept of 'vegetation series'. In: Geographical changes in vegetation and plant functional types. Springer, pp 3–18

Carrión JS, Fernández S (2009) The survival of the 'natural potential vegetation'concept (or the power of tradition). J Biogeogr 36:2202–2203

Chiarucci A, Araújo MB, Decocq G et al (2010) The concept of potential natural vegetation: an epitaph? J Veg Sci 21:1172–1178

Demissew S, Cribb P, Rasmussen F (2004) Field guide to Ethiopian Orchids. Royal Botanical Garden, Kew

Demissew S, Nordal I, Stabbetorp OE (1999) Endemism and patterns of distribution of the genus Aloe (Aloaceae) in the flora of Ethiopia and Eritrea. In: Friis I, Ryding O (eds) Biodiversity research in the horn of Africa region: Proceeding of the third international symposium on the flora of Ethiopia and Eritrea at the Carlsberg Academy, Copenhagen, August 25–27. The Royal Danish Academy of Sciences and Letters, Copenhagen, Denmark

Demissew S, Wondafrash M, Dellellegn Y (1996) Ethiopia's natural resources base. In: Edwards S (ed) Important bird areas of Ethiopia: a first inventory. Ethiopian Wildlife and Natural History Society, Addis Ababa, Ethiopia

Egziabher TBG (1988) Vegetation and environment of the mountains of Ethiopia: implications for utilization and conservation. Mt Res Dev 8:211–216

Friis I (1992) Forests and forest trees of Northeast Tropical Africa. HMSO, London, UK

Friis I, Demissew S (2001) Vegetation maps of Ethiopia and Eritrea. A review of existing maps and the need for a new map for the Flora of Ethiopia and Eritrea. Biol Skr 54:399–439

Friis I, Demissew S, van Breugel P (2010) Atlas of the potential vegetation of Ethiopia. The Royal Danish Academy of Sciences and Letters, Copenhagen, Denmark

Gebrehiwot K (2020) Vegetation ecology and dynamics of land cover change of Abune Yosef mountain range, Lasta district, Amhara regional state, northern Ethiopia. Addis Ababa University

Gebrehiwot K, Woldu Z, Fekadu M, et al (2019) Classification and ordination of plant communities in Abune Yosef mountain range, Ethiopia. Acta Ecol Sin. https://doi.org/10.1016/j.chnaes.2019.12.001

Hedberg O (1951) Vegetation belts of East African mountains. Sven Bot Tidakrift 45:140–195

Hedberg O (1964) Features of Afroalpine plant ecology. Almqvist & Wiksells Boktryckeri AB, Uppsala

Hedberg O (1969) Evolution and speciation in a tropical high mountain flora. Biol J Linn Soc 1:135–148

Hengl T, Walsh MG, Sanderman J et al (2018) Global mapping of potential natural vegetation: an assessment of machine learning algorithms for estimating land potential. Peer J 6:e5457. https://doi.org/10.7717/peerj.5457

Pichi-Sermolli RE (1957) Una carta geobotanica dell' Africa orientale (Eritrea, Etiopia, Somalia). Webbia 13:15–132

Somodi I, Ewald J, Ákos B-F, Molnár Z (2021) The relevance of the concept of potential natural vegetation in the Anthropocene. Plant Ecol Divers 14:13–22. https://doi.org/10.1080/17550874.2021.1984600

Tuxen R (1956) Die huetige potentielle naturliche Vegetation als Gegestand der Vegetation-skarierung. Angew Pflanzensozioligie 13:5–42

van Breugel P, Kindt R, Lillesø JP, et al (2015) Potential Natural Vegetation Map of Eastern Africa (Burundi, Ethiopia, Kenya, Malawi, Rwanda, Tanzania, Uganda and Zambia). Version 2.0. In: For. Landsc. Denmark World Agrofor. Cent.

White F (1983) The vegetation of Africa. A descriptive memoir to accompany the UNESCO/AETFAT/UNESCO vegetation map of Africa. UNESCO, Paris, France

Woldu Z (1999) Forest in the vegetation types of Ethiopia and their status in the geographical context. In: Edwards S, Demissie A, Bekelle Y, Haase G (eds) forest genetic resources- conservation: principles strategies and action. Addis Ababa, Ethiopia

Zerbe S (1998) Potential natural vegetation: validity and applicability in landscape planning and nature conservation. Appl Veg Sci 1:165–172

Chapter 4
Plant Biodiversity Conservation and Lost Opportunities in Ethiopia

Abstract Ethiopia is endowed with heterogeneous landscapes. The diverse topography has a great impact on the spatial rainfall and temperature variability across the country. In this diverse topography, about 14 floristic regions are recognized in Ethiopia. A synthesis analysis of the Flora of Ethiopia and Eritrea revealed that there are above 5757 higher plants of which about 10% are endemic. The endemic plants reported from the flora area comprise 17 and 10% of the endemic and near-endemic species richness of the Horn of Africa. However, several new species to the flora of Ethiopia have been published after the completion of the flora project. Thus, the ongoing recording of new species elucidates that the number of plants reported in the flora of Ethiopia is by no means complete. Future protected areas need to represent the full variety of habitat types, species assemblages, ecological processes, or other natural features that are characteristic of a given region. In Ethiopia, protected areas are fairly representative though the lowlands are poorly represented. However, the irreplaceability and vulnerability criteria had been poorly considered during the establishment of the most protected areas.

Keywords Conservation approach · Flora of Ethiopia · Floristic regions · Representativeness · Vulnerability/irreplaceability · *Taverniera abyssinica*

4.1 Introduction

Ethiopia is endowed with heterogeneous landscapes. The elevation ranges from the lowest in the world (126 m below sea level) in the Danakil depression to about 4533 m above sea level (m a.s.l.) in Ras Dejen Mountain which is part of the Simien Mountains. The diverse topography has a great impact on the spatial rainfall and temperature variability across the country. These determine the vegetation/ecosystem types, which further determine the floristic regions across the country. Human-induced activities such as land-use changes also play a crucial role in determining plant biodiversity conservation in the floristic regions of Ethiopia.

K. G. Yaynemsa, *Plant Biodiversity Conservation in Ethiopia*,
https://doi.org/10.1007/978-3-031-20225-4_4

4.1.1 Ethiopian Floristic Regions

The surface area of Ethiopia is about 1.12 million square kilometers. The Flora of Ethiopia and Eritrea (FEE) project recognized about 14 floristic regions in Ethiopia. Before Eritrea declared its independence, the flora of Eritrea (two floristic regions) was included as part of the flora of Ethiopia and Eritrea. The administrative provinces of imperial Ethiopia were the important criteria in categorizing the floristic regions. As depicted in Fig. 4.1, the Ethiopian floristic regions are Afar (AF), Tigray (TU), Gondar (GD), Welo (WU), Gojam (GJ), Shewa (SU) Arsi (AR), Bale (BA), Sidamo (SD), Gamo Gofa (GG), Kefa (KF), Harerge (HA), Illubabor (IL), and Welega (WL) floristic regions. Recently, however, a hierarchical clustering approach, which is different from the administrative province's system was developed (Wang et al. 2020). This study identified two floristic sub-provinces, five floristic districts, and 13 floristic sub-districts, as well as three centers of species endemism associated with three floristic sub-districts, were identified.

These floristic regions (both provincial and hierarchical clustering) are by no means identical to the current administrative boundaries (regional states) of the country. The hierarchical clustering approach is more similar to the potential natural vegetation (Chap. 3). Even though there are overlaps in floristic composition, the floristic regions have their distinct features. The following subsection discusses the floristic regions.

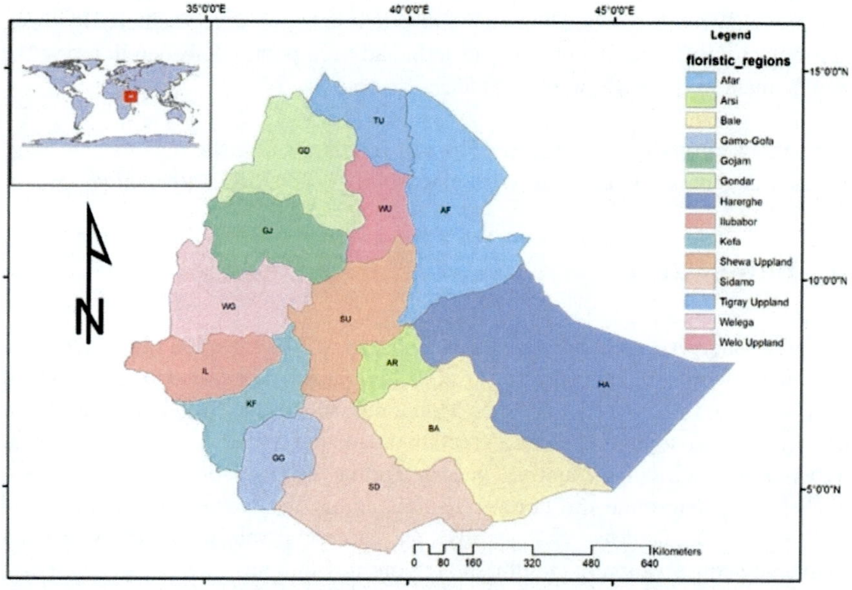

Fig. 4.1 Floristic regions of Ethiopia

4.1.1.1 Afar (AF) Floristic Region

This floristic region represents the eastern and northeastern lowlands of Ethiopia. The Eastern Eritrea, Djibouti, Tigray, Welo, Shewa, Arsi, and Hararge floristic regions border it. The vegetation types found in this floristic region are dominantly desert and semi-desert scrubland (DSS) and *Acacia-Commiphora* woodland (Chap. 3). This floristic region is known for its extremely high temperature and low rainfall. It harbors the lowest species richness, endemic, and near-endemic with about 200 taxa and 25 taxa, respectively (Friis et al. 2001). The inaccessibility of the floristic region during the flora exploration project might have caused under-exploration, which in return contributed to the low species richness. Even though the recurrent drought and anthropogenic disturbances could have some impact on species richness in this floristic region, a complete survey of the floristic region might likely add new taxa that were not previously reported. The new record (*Tragus mongolorum* Ohwi) to Afar floristic region as well as FEE documented by Friis et al. (2011) after the survey in 2007 supports the earlier assumption.

4.1.1.2 Tigray (TU) Floristic Region

East and West Eritrea, Afar, Gondar, and Welo floristic regions border Tigray floristic region. It is the fourth species-rich floristic region in Ethiopia following Sidamo, Shewa, and Hararge. Concerning the endemic and near-endemics, however, it is ranked sixth (Friis et al. 2001, 2005). However, a recent report lowered the rank to seventh (Demissew et al. 2021). Recent studies reveal that there are new records to this floristic region. For instance, Hishe (2019) documented that *Phyllanthus mooneyi* M. G. Gilbert is a new record to Tigray floristic region. Furthermore, 17 species (*Acacia nilotica* (L.) Willd. ex Delile., *Ageratum conyzoides* Sieber ex Steud., *Antherotoma naudinii* Hook.f., *Brassica carinata* A. Braun, *Buddleja cordata* Kunth, *Commelina benghalensis* Forssk., *Desmodium repandum* (Vahl) Poir., *Elaeodendron buchananii* (Loes.) Loes., *Eugenia bukobensis* Engl., *Galiniera saxifraga* (Hochst.) Bridson, *Hyparrhenia cymbaria* (L.) Stapf., *Sida cordifolia* Forssk., *Tagetes minuta* L., *Terminalia laxiflora* Engl., *Terminalia macroptera* Guill. and Perr., *Trachyspermum ammi* (L.) Sprague, and *Ziziphus mucronate* Willd.) were reported as new records to the floristic region (Girmay 2021). This shows that an extensive survey of the floristic regions, in general, could change the species richness reported by the flora project. Nevertheless, the national species richness might not be changed unless a new species that was not included in the FEE is recorded.

4.1.1.3 Gondar (GD) Floristic Region

The floristic region has several species shared with the adjacent floristic regions. This could be due to its geographical proximity. It is bordered by Tigray, Welo, and Gojam floristic regions Furthermore, Sudan also borders the floristic region. It is ranked sixth in its taxonomic diversity whereas fifth in endemic and near-endemics (Friis et al. 2001; Demissew et al. 2021) although ranked fifth in another study (Friis et al. 2005). The highest elevation in Ethiopia is found in this floristic region, Simien Mountain. Like the previous floristic regions, new records are reported in this floristic region. For instance, *Leucas aequistylosa* Sebald., *Parthenium hysterophorus* L., and *Calotropis procera* (Aiton) W. T. Aiton (Belay 2018), and *Hyparrhenia diplandra* (Hack.) Stapf are reported as new records in this floristic region *P. hysterophorus* and *C. procera* are invasive and potentially invasive species that are rampantly distributed almost in all floristic regions.

4.1.1.4 Welo (WU) Floristic Region

This floristic region is the second species-poor after Afar floristic region (Friis et al. 2001, 2005; Demissew et al. 2021). However, a recent study documented 93 species that were not recorded in this floristic region (Gebrehiwot et al. 2019), which might improve the species richness rank. Hence, a robust survey, particularly in areas that were not accessible in the flora project is commendable.

4.1.1.5 Gojam (GJ) Floristic Region

Gojam floristic region is ranked ninth in species richness and seventh in endemic and near-endemic species richness (Friis et al. 2001, 2005) although it is down to 10th in a recent study (Demissew et al. 2021). This floristic region is bordered by Gondar, Shewa, Welo, and Wolega floristic regions. It is also bordered by Sudan. *Indigofera bracteolata* DC. and *Clerodendrum triflorum* Vis. are some of the new species recorded that are not included in the flora of Ethiopia (Friis et al. 2011). Few researchers documented new species recorded in Gojam floristic region. For example, *Inula confertiflora* A. Rich., *Mikaniopsis clematoides* (Sch.Bip. ex A.Rich.) Milne-Redh., *Senecio myriocephalus* Sch.Bip. ex A.Rich., and *Leucas abyssinica* (Benth.) Briq. were recorded as new records to Gojam floristic region (Achiso 2014). Furthermore, *Carum piovanii* Chiov., *Crotalaria rosenii* (Pax) Milne-Redh. ex Polhill, *Justicia diclipteroides* subsp. *aethiopica* Hedrén, *Kalanchoe densiflora* var. *subpilosa* Cufod., *Maytenus addat* (Loes.) Sebsebe, *Otostegia tomentosa* subsp. *tomentosa* A.Rich., *Polystachya caduca* Rchp.f., and *Tiliacora troupinii* Cufod. were also recorded as new records to the floristic region (Berhanu 2018). Considering these new records, the floristic region's taxonomic diversity might be underestimated.

4.1.1.6 Shewa (SU) Floristic Region

Shewa floristic region is the second species-rich following Sidamo floristic region (Friis and Demissew 2001; Friis et al. 2001, 2005; Demissew et al. 2021). It is one of the three floristic regions with above one thousand taxa. This floristic region is bordered by Gojam, Welo, Afar, Wolega, Kefa, and Arsi floristic regions. With about 200 taxa, it's ranked fourth in endemic and near-endemics taxa (Friis et al. 2001). However, with the new species recorded in this floristic region, the number of taxa might slightly increase. For instance, about 13 new species (e.g., *Conostomium quadrangulare* (Rendle) Cufod., *Cyperus conglomeratus* Rottb., *Isodon ramosissimus* (Hook.f.) Codd, *Sedum glomerifolium* M. G. Gilbert) were recorded in Shewa floristic region (Tilahun 2018). This floristic region is part of the Capital, Addis Ababa. Thus, we can assume this floristic region is the most accessible floristic region in the country. If 13 species are documented as new records in a single study in the most accessible floristic region, this implies there is a lot to be done. With these additional taxa, the taxonomic diversity of the floristic region would slightly increase.

4.1.1.7 Arsi (AR) Floristic Region

Arsi floristic region is dominantly found in the rift valley though there are high mountains to the eastern flank (e.g., Arsi Mountain). This floristic region is one of the species-poor floristic regions in Ethiopia. Furthermore, the endemic and near-endemics are also poor in AR (Friis et al. 2001). Afar, Hararge, Shewa, and Bale floristic regions bordered this floristic region.

4.1.1.8 Bale (BA) Floristic Region

This floristic region is found in the southeastern part of Ethiopia. It is surrounded by Arsi, Hararge, and Sidamo floristic regions of Ethiopia and Somalia as well. Even though it was ranked seventh in species richness, the endemic and near-endemic species richness is high (Friis et al. 2001, 2005). However, the latest study on the Flora of Ethiopia and Eritrea improved the rank to fourth (Demissew et al. 2021). However, the new species recorded in Bale floristic region after the publication of the FEE may slightly increase the species richness. For instance, *Leucas gypsicola* A. J. Paton, Friis and Sebsebe (Paton et al. 2018), *Croton elkerensis* Friis and M. G. Gilbert, *Gnidia elkerensis* Friis and Sebsebe, and *Plectranthus spananthus* A. J. Paton, Friis and Sebsebe (Friis et al. 2018) were recorded as new species to the floristic region and Ethiopia as well. In addition, an endemic and endangered species, *Barleria gidoleensis* Ensermu and I. Darbysh. that was not included in the flora books is recently recorded in this floristic region (Kelbessa and Darbyshire 2018). This implies that extensive floristic exploration could increase the taxonomic diversity of this floristic region and Ethiopia as well.

4.1.1.9 Sidamo (SD) Floristic Region

This floristic region is top-ranked both in taxonomic diversity, and endemic and near-endemic richness (Friis et al. 2001, 2005; Wang et al. 2020; Demissew et al. 2021). Sidamo floristic region is bordered by Bale, Arsi, and Gamo Gofa floristic regions in Ethiopia and Kenya in East Africa. An extensive survey on the floristic region could slightly increase the taxonomic richness. Several researchers documented new species records in this floristic region. New species that were not included in the Sidamo floristic region such as *Rhus retinorrhoea* Steud. ex Oliv., *Dracaena afromontana* Mildbr., *Erythrococca trichogyne* (Müll.Arg.) Prain, *Sida tenuicarpa* Vollesen and *Rubus volkensii* Engl. were recently documented in this floristic region (Kebede et al. 2013). Besides, Friis et al. (2011) recorded *H. diplandra* as new record to this Sidamo floristic region. Another study documented 16 species (e.g., *Agarista salicifolia* (Lam.) G.Don, *Halleria lucida* L., *Ilex mitis* (L.) Radlk.) new record to Sidamo floristic region (Woldearegay 2017). Furthermore, an endemic and endangered species, *B. gidoleensis* that was not included in the flora books is recently recorded in this floristic region (Kelbessa and Darbyshire 2018).

4.1.1.10 Gamo Gofa (GG) Floristic Region

The floristic regions Kefa and Sidamo surrounded this floristic region. It is also bordered by Kenya. Even though the taxonomic diversity is medium, the endemic and near-endemic species richness is poor (< 80), which is in the same range as Arsi, Gamo Gofa, Welega, and Welo floristic regions (Friis et al. 2001). Some new species that were not included in the Gojam floristic for instance *I. bracteolata* were recorded (Friis et al. 2011). The endemic, *Vepris dainellii* (Pic.Serm.) Kokwaro, that was not included in this floristic region in the FEE is reported as a new record for GG (Denu and Desissa 2013). An endemic and endangered species *B. gidoleensis*, that was not included in the flora books is recently recorded in this floristic region (Kelbessa and Darbyshire 2018).

4.1.1.11 Kefa (KF) Floristic Region

Kefa floristic region is found in southwestern Ethiopia, where the remnant intact forests are found. Furthermore, it is the origin of *Coffea arabica* L. This floristic region is bordered by Gamo Gofa, Shewa, Wolega, and Illubabor floristic regions. Moreover, Sudan also bordered this floristic region. Even though the endemic and near-endemic species richness is relatively low, the taxonomic diversity is high (Friis et al. 2001). For instance, *H. diplandra* was recorded in this floristic region as a new species record (Friis et al. 2011). Recently, about 19 species (e.g., *Alchemilla pedate* Hochst. ex A. Rich., *Acanthopale ethio-germanica* Ensermu, *Oliverella hildebrandtii (Engl.) Tiegh.*) that were not included in the Kefa floristic region in the FEE were reported as new records (Addi 2018). An endemic and Near Threatened (NT) species,

Barleria baluganii Ensermu, which was not included in the FEE is recently reported in this floristic region (Kelbessa and Darbyshire 2018).

4.1.1.12 Harerge (HA) Floristic Region

This species-rich floristic region is distributed in eastern Ethiopia, which is dominantly a lowland. The taxonomic diversity and the endemic and near-endemic richness are estimated to be above 1100 and 300 respectively (Friis et al. 2001). Due to its inaccessibility, however, the survey could have underestimated it. Nevertheless, some researchers are documenting new records in this floristic region. For instance, an endemic and endangered (EN) species, *Barleria ferox* Ensermu and I. Darbysh. which was not included in the FEE is recently reported in this floristic region (Kelbessa and Darbyshire 2018).

4.1.1.13 Illubabor (IL) Floristic Region

Illubabor floristic region is bordered by Kefa and Wolega floristic regions in Ethiopia, and Sudan in east Africa. It comprises relatively low taxonomic diversity, and endemic and near-endemic species richness (Friis et al. 2001). After the completion of the flora project, however, reports of new species records emerged. Fifty-seven species were reported as new records to this floristic region though the species collected may have some overlap with Kefa floristic region (Kassa 2017). Even though it is difficult to compare the species list with that of Kassa (2017), Kefalew (2019) also reported 59 new species to this floristic region. An endemic and Near Threatened (NT) species, *B. baluganii,* which was not included in the FEE is recently reported in this floristic region (Kelbessa and Darbyshire 2018).

4.1.1.14 Welega (WL) Floristic Region

At the regional level, this floristic region is also bordered by Sudan. On the other hand, it is surrounded by Shewa, Illubabor, Kefa, and Gojam floristic regions. It comprises relatively low taxonomic diversity, and endemic and near-endemic taxonomic richness (Friis et al. 2001). Few studies reported new species records after the publication of FEE. New species records such as *Phyllanthus chevalieri* Beille., *Syngonanthus wahlbergii* (Wikstr. ex Korn.) Ruhland, *H. diplandra,* and *Wahlenbergia paludicola* Thulin were reported in this floristic region (Friis et al. 2011). Thirty-one new species to Wolega floristic region that were not included in the FEE reported were reported (Gurmessa et al. 2013). About 95 new species recorded to Welega floristic region was also reported (Balemie 2019). Besides, an endemic and Near Threatened (NT) species, *B. baluganii,* which was not included in the FEE is recently reported in this floristic region (Kelbessa and Darbyshire 2018).

4.2 Plant Conservation Status

Ethiopia is one of the 36 biodiversity hotspots (Mittermeier et al. 2004, 2011; Hoffman et al. 2016). Biodiversity hotspots are biogeographic areas around the world that are a significant reservoir of biodiversity as well as threatened by anthropogenic activities. The topographic heterogeneity has a significant contribution to plant diversity although the rampantly undergoing human activities such as land use and land cover changes are exerting threats to biodiversity.

The number of plant species in Ethiopia reported varies. Despite the variation in the number of plants reported, Ethiopia hosts the fifth-largest floral diversity in tropical Africa (Maczulak 2010). A synthesis analysis of the Flora of Ethiopia and Eritrea revealed that there are above 5757 higher plants of which about 10% are endemic (Kelbessa and Demissew 2014). The endemic plants reported from the flora area comprise 17 and 10% of the endemic and near-endemic species richness of the Horn of Africa (Friis et al. 2005). However, several new species to the flora of Ethiopia have been published after the completion of the flora project (Table 4.1). Thus, the ongoing recording of new species reveals that the number of vascular plants reported in the flora of Ethiopia is by no means complete. Besides, the number of vascular plants is beyond 6000. Although it might not make significant difference, the number of endemic plant species could also be higher than reported (Table 4.1).

Plants have myriad economic, ecological, and cultural importance. Consequently, they are exploited for these services. Exploitation, habitat destruction, and fragmentation, invasive species, etc., are some examples of the threats plants are facing in the Anthropocene. The man-induced threats often determine the conservation status of plant biodiversity. There are few approaches employed to assess the threat status of biodiversity at different spatial scales. The International Union for Conservation of Nature (IUCN) Red List criteria is one of the widely used approaches. However, rarity indices are also used to determine the conservation status of an organism.

Out of the 310, 343 described plant (including green and red algae, and mosses) species, 24,230 (8% of the described plant species) have been evaluated by the International Union for Conservation of Nature (IUCN). The threatened plant species increased from 5328 in 1996/98 to 12,505 in 2017 (IUCN 2017). This two-decade data show plants are at an ultimate threat. Considering, the number of species assessed and the remaining unassessed and undescribed species, the global coverage of floristics, this IUCN work is the tip of the iceberg. Furthermore, regional, national, and local evaluations are very limited.

According to the world conservation union report IUCN (1998), 163 (2.5%) of the vascular plants in Ethiopia were threatened. IUCN Red List of endemic trees and shrubs of Ethiopia and Eritrea assessed 135 endemic plants (31 trees and 104 shrubs) and revealed that 116 of them are threatened (Vivero et al. 2005). However, assessment of endemic flowering taxa assessed 596 taxa covering 550 species which are 9.7% of the estimated plants in FEE (Vivero et al. 2006) and found 77.4% of them are threatened.

Table 4.1 New species published after the publication of FEE

No	Species	Family	Sources
1	*Phyllanthus chevalieri* Beille	Euphorbiaceae	Friis et al. (2011)
2	*Indigofera bracteolata* DC	Fabaceae	
3	*Wahlenbergia paludicola* Thulin	Campanulaceae	
4	*Clerodendrum triflorum* Vis	Lamiaceae	
5	Tragus mongolorum Ohwi	Poaceae	
6	*Hyparrhenia diplandra* var. *mutica* (W.D. Clayton) T. Cope	Poaceae	
7	*Syngonanthus wahlbergii* (Wikstr. ex Korn.) Ruhland	Eriocaulaceae	
8	*Elaeis guineensis* Jacq	Arecaceae	
9	*Croton elkerensis* Friis and M.G.Gilbert	Euphorbiaceae	Friis et al. (2018)
10	*Gnidia elkerensis* Friis and Sebsebe	Thymelaeaceae	
11	*Plectranthus spananthus* A. J. Paton, Friis and Sebsebe	Lamiaceae	
12	*Leucas gypsicola* A. J. Paton, Friis and Sebsebe	Lamiaceae	Paton et al. (2018)
13	*Euphorbia venefica* Trémaux ex Kotschy	Euphorbiaceae	Weber et al. (2020)
14	*Kalanchoe hypseloleuce* Friis and M. G. Gilbert	Crassulaceae	Friis et al. (2017)
15	*Commicarpus macrothamnus* Friis and O. Weber sp. nov	Nyctaginaceae	Friis et al. (2016)
16	*Commicarpus leleensis* Friis and Sebsebe sp. nov	Nyctaginaceae	
17	*Alchornea cordifolia* (Schumach. and Thonn.) Muell. Arg	Euphorbiaceae	Friis and Harris (2013)
18	*Aloe benishangulana* Sebsebe and Tesfaye	Aloaceae	Friis and Harris (2013)
19	*Aloe ghibensis* Sebsebe and Friis	Aloaceae	
20	*Aloe weloensis* Sebsebe	Aloaceae	
21	*Aloe welmelensis* Sebsebe and Nordal	Aloaceae	
22	*Barleria gidoleensis* Ensermu and I. Darbysh	Acanthaceae	Kelbessa and Darbyshire (2018)
23	*Barleria baluganii* Ensermu	Acanthaceae	
24	*Barleria ferox* Ensermu and I. Darbysh	Acanthaceae	
25	*Barleria negeleensis* Ensermu and I. Darbysh	Acanthaceae	

(continued)

Table 4.1 (continued)

No	Species	Family	Sources
26	*Zygotritonia benishangula* Goldblatt and Sebsebe	Iridaceae	Friis and Harris (2013)
27	*Commicarpus ogadenensis* Thulin	Nyctaginaceae	Thulin (2021)
28	*Blepharis gypsophila* Thulin and Vollesen	Acanthaceae	Thulin and Vollesen (2015)
29	*Caralluma wilhelmii* Thulin	Apocynaceae	Thulin (2009a)
30	*Ceropegia gypsophila* Thulin	Apocynaceae	
31	*Ceropegia kaariyei* Thulin	Apocynaceae	
32	*Ceropegia marronina* Bruyns and Hanáček	Apocynaceae	http://powo.science.kew.org
33	*Ceropegia pseudorhynchantha* Bruyns	Apocynaceae	http://powo.science.kew.org
34	*Chlorophytum mamillatum* Elden and Nordal	Asparagaceae	http://powo.science.kew.org
35	*Cibirhiza spiculata* Thulin and Goyder	Apocynaceae	Thulin et al. (2008)
36	*Coccinia ogadensis* Thulin	*Cucurbitaceae*	http://powo.science.kew.org
37	*Euphorbia bertemariae* Bisseret and Dioli	Euphorbiaceae	Bisseret and Dioli (2005)
38	*Euphorbia gilbertiana* Bisseret and Specks	Euphorbiaceae	Bisseret and Specks (2006)
39	*Euphorbia kabridarensis* Thulin	Euphorbiaceae	Thulin (2009b)
40	*Euphorbia michaelii* Thulin		
41	*Euphorbia piceoides* Thulin		
42	*Euphorbia suborbicularis* Thulin		
43	*Huernia delicata* Plowes	Apocynaceae	http://powo.science.kew.org
44	*Huernia sudanensis* Plowes	Apocynaceae	http://powo.science.kew.org
45	*Hymenasplenium kenyense* Li Bing Zhang, K. W. Xu and Kamau	Aspleniaceae	http://powo.science.kew.org
46	***Kleinia ogadensis* Thulin**	Asteraceae	http://powo.science.kew.org
47	***Lepidagathis pseudoaristata* Ensermu**	Acanthaceae	Kelbessa (2009)
48	*Acanthopale aethio-germanica* Ensermu	Acanthaceae	
59	*Rhinacanthus mucronatus* Ensermu	Acanthaceae	
50	***Momordica argillicola* Thulin**	Cucurbitaceae	http://powo.science.kew.org
51	*Mucuna mooneyi* T. M. Moura, Gereau and G. P. Lewis	Fabaceae	de Moura et al. (2015)

(continued)

Table 4.1 (continued)

No	Species	Family	Sources
52	*Polystachya sosefii* Baranow and Mytnik	Orchidaceae	Baranow and Mytnik (2009)
53	*Pterodiscus somanei* Peckover	Pedaliaceae	Peckover (2020)
54	*Scleria pseudohispidior* Bauters	Cyperaceae	Bauters et al. (2018)
55	*Seddera ogadenensis* Sebsebe	Convolvulaceae	Demissew and Mill (2009)
56	*Seddera rhodantha* R. R. Mill		
57	*Solanum pronum* Thulin	Solanaceae	Thulin (2009c)

Moreover, a recent study reported two Endangered (EN), one Near Threatened (NT), and one Vulnerable (VU) (Kelbessa and Darbyshire 2018). An assessment of the Aloe species of the Horn of Africa revealed that 39% of the 88 *Aloe* species are threatened (Bachman et al. 2020). Other local rarity assessments in the Ethiopian mountain range revealed nearly 90% of the plants assessed are rare (Gebrehiwot et al. under review). A recent global report on the status of tree species revealed that 44 (7%) of tree species are threatened (BGCI 2021). It is not only the wild plants but some hardly known crops for instance *Ensete ventricosum* are also threatened (Antonelli et al. 2020).

What makes it worst is, the more conservation status assessment is undergone the more the threatened plants reported. Hence, to understand the conservation status of the plant biodiversity and gear into evidence-based conservation (Chap. 10) a comprehensive assessment is required. The approach could be either IUCN or other complementary approaches.

4.3 Plant Conservation Approaches Implemented in Ethiopia

Globally, there are different approaches of conservation planning and prioritization. The most common approaches practiced are species, site level, and land/seascape level conservation planning and prioritization. Each conservation planning and prioritization approaches have pros and cons. Hence, policy- and decision-makers must consider this before the conservation project implementation. The species-, site- and landscape-level approaches are briefly discussed in the following subsections.

4.3.1 Species-Level Conservation

Species is the fundamental component in biodiversity conservation. This approach is the oldest and well-known approach globally. The primary target of this conservation planning and prioritization approach is either single or multiple species. In order to apply this approach, a taxonomic description and conservation status of a species are required, which is ideally possible and practically not manageable. Most plants are taxonomically described. Nevertheless, about 10% are assessed for the IUCN threat status both at the global and national levels. Furthermore, it needs manpower with taxonomy and IUCN Red List criteria expertise. Hence, this makes it difficult to favor this approach. If these challenges are overcome, however, this approach could have a crucial role in preventing species extinction.

From the plant conservation perspective, this approach has a short history in Ethiopia. It is often challenging to get documents on this topic. Nonetheless, it is possible to say the species-level conservation approach widely started after the establishment of the Gullele Botanical Garden. Its establishment motivated the conservation of threatened plants ex situ, by relocating them to this botanical garden. However, its scope is extremely limited.

4.3.2 Site-Level Conservation

This approach has a long history than the species-level conservation approach. Natural reserve establishment started about 600 years ago in Ethiopia (Pankhurst 1989) despite most of the sites being established from the 1960s onwards. The national parks and National Forest Priority Areas are examples of site-level conservation approaches.

Habitat destruction or transformation is the major driver that threatens nearly all species. Thus, conserving habitat *sense stricto* conserves the species found in that habitat. Stakeholders might invest less in species taxonomic description and IUCN conservation status assessments. Moreover, relative to the species-level conservation approach, this approach needs less trained manpower.

However, this approach has its drawbacks. These include the challenge to identify biodiversity areas that have national, regional, and global importance and how to priorities the most urgent for conservation. In addition, in human-dominated and fragmented landscapes, the site-level conservation approach might not be successful as it is supposed to be. Because, if the ecological networks are disconnected, dispersal and pollination may be hampered. To alleviate these constraints, policy- and decision-makers may opt for landscape-level conservation.

4.3.3 Landscape-Level Conservation

Species and site-level conservation approaches have their own merits from different perspectives. However, they will never be sufficient. The drawbacks were stated in the above sections. For instance, it would be difficult to implement either species or site-level conservation approach in human-dominated landscapes. Hence, thinking of larger to benefit biodiversity conservation is commendable. This includes establishing habitat corridors to connect the fragmented sites along the landscape so that dispersal and pollination across the landscape are improved. This strengthens the ecological networks of the landscape.

However, this approach is criticized for some obvious reasons such as it could divert resources from high priorities to low priority areas, could spread diseases along the landscape, and pave the way for invasion. Furthermore, organizing supporting data about the spatial scale required for the species in the landscape could be tiresome and resource-consuming.

Therefore, decision- and policy-makers, as well as conservation practitioners, should consider the pros and cons of the three-conservation planning and prioritization approaches before implementing conservation any projects.

4.4 Major Threats to Plant Biodiversity in Ethiopia

The major threats to biodiversity are more or less similar across the continents. The magnitude of the particular threat, nonetheless, could not be similar. For instance, livestock grazing, instability, recurrent droughts, forest fires, settlements, and agricultural encroachment to forests were reported as the main threats to the *Acacia-Commiphora* woodland and bushland vegetation in the Bale Mountains, Ethiopia (Vivero et al. 2006). A threat assessment on *Aloe* species in the horn of Africa showed that farming, unintentional logging, and wood collection are the main threats to *Aloe* species (Bachman et al. 2020). Furthermore, a recent study that documented a new species to the flora of Ethiopia implied that habitat loss, expanding plantation agriculture, exotic tree forestry, timber extraction, settlement, agricultural expansions, and population growth are the threats for the species documented (Kelbessa and Darbyshire 2018). Although it might not be the direct cause, the ever-increasing human population is the main threat to plant biodiversity. As the human population increases the need for resources linearly grows. As a result, human beings will be forced to exploit the resources to satisfy their demands, which in return threatens plant biodiversity. Hence, managing human population growth and promoting sustainable utilization of resources could reduce the threats to plant biodiversity.

4.5 Lost Opportunity in Plant Biodiversity Conservation

It is believed that protected areas play a crucial role in biodiversity conservation. Protected area cover is also the crux in biodiversity conservation. According to the fundamental species richness-area principle, the larger the protected area the high species richness it hosts. Sometimes, small reserves could also harbor significantly higher species richness (Chap. 5).

Ethiopia met the Target 11 of the Aichi agreement (Bacon et al. 2019). However, the post-2020 target 3 proposes that "ensure that at least 30% of the globally land areas and of sea areas, especially areas of particular importance for biodiversity and its contribution to people are conserved effectively and equitable managed, ecologically representative and well-connected systems of protected areas and other effective area-based conservation measures (OECMs) and integrated into the wider landscape and seascape" (CBD 2020).

What matters most is, however, how effective the established protected areas are. Are they representative? Do they consider vulnerability and irreplaceability? The scope of these questions is broad. Irreplaceability is often associated with ecological uniqueness or endemism and its importance in contributing to a specific set of conservation targets. For instance, if a species is endemic to a particular place only and its habitat is lost, it could not be replaced. Vulnerability is a threat of loss of a habitat or species. Whereas, representativeness is the need for protected areas to represent the full variety of habitat types, species assemblages, ecological processes, or other natural features that are characteristic for a given region.

In Ethiopia, protected areas are fairly representative though the lowlands (e.g., the desert and semi-desert scrubland) are poorly represented. However, the irreplaceability and vulnerability criteria had been poorly considered during the establishment of the most protected areas. Some national parks in Ethiopia are objectively established to protect a particular species (often animals). For instance, the Simien mountains National Park was established to conserve the endemic and threatened Walia Ibex (*Capra walie*). Nevertheless, it is barely noticed that protected areas are established by considering irreplaceability and vulnerability for plants. For example, Bachman et al. (2020) showed that the protected areas in Ethiopia offer limited coverage for the most threatened and endemic *Aloe* species, and the most recently designated protected areas do not overlap with *Aloe* distributions (Fig. 4.2).

Ninety-four geographic coordinates of thirteen threatened species (Critically endangered, Endangered, and Vulnerable) were collected from GBIF (https://www.gbif.org/). These points were overlaid over the map of Ethiopia to see if these species are appropriately represented in the already protected areas (Fig. 4.3). Surprisingly, only about 10% of the geographic coordinates fall within the protected areas. One of the most cited endemic medicinal plants and critically endangered (CR) species, *Taverniera abyssinica* A. Rich., is represented in none of the 138 protected areas in Ethiopia. Thus, this reveals that the vulnerable and irreplaceable species might not be sufficiently represented in the Ethiopian protected areas. This is without considering the species range shift (Chap. 1) happening globally due to global warming. Policy- and decision-makers as well as conservation practitioners, hence, need to consider the vulnerability and irreplaceability of the plants in the era of global warming prior to establishing conservation reserves and improving ecological connectivity (Chap. 9).

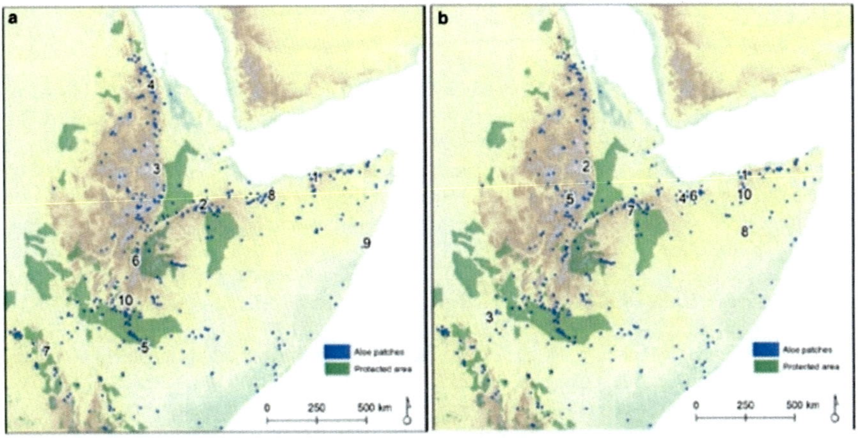

Fig. 4.2 Maps illustrating the location of the top ten sites in order of priority (1 is the highest priority) to capture all species (**a**), and all threatened species (**b**), with the minimum amount of additional PA. All sites are shown as blue patches and the existing PA network is shown with green polygons. Adopted from (Bachman et al. 2020)

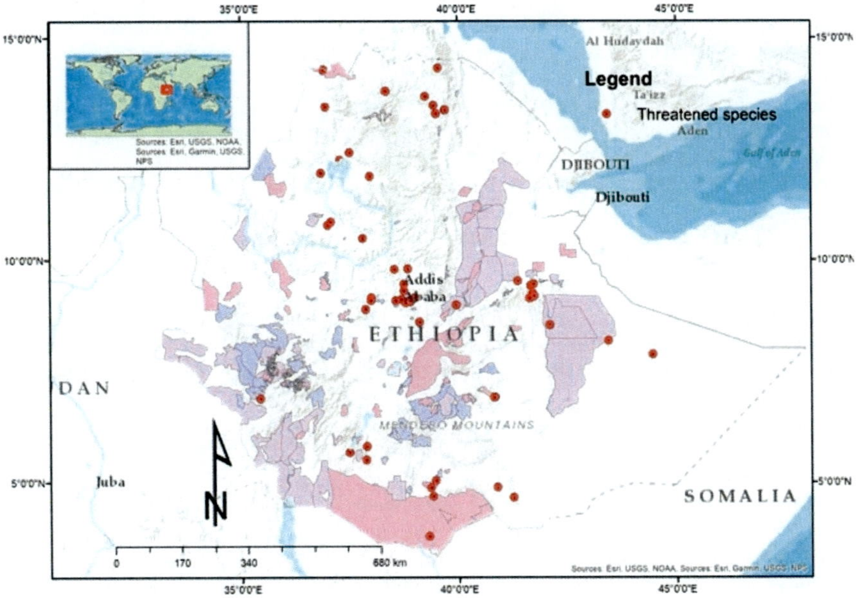

Fig. 4.3 Red dots are the GPS coordinates of randomly selected thirteen threatened, most of them endemic species. Whereas the polygons are some forms of protected areas in Ethiopia

References

Achiso Z (2014) The distribution of the Woody Vegetation along altitudinal ranges from Abay (Blue Nile) Gorge to Choke Mountain, East Gojjam Zone of Amhara Regional State. Addis Ababa University, North West Ethiopia

Addi A (2018) The ecology, carbon stock, bee forage diversity in a moist Afromontane forest of Gesha and Sayilem districts in Kaffa zone, south west Ethiopia. Addis Ababa University

Antonelli A, Fry C, Smith R., et al (2020) State of the world's plants and fungi 2020. Royal Botanic Gardens, Kew, UK

Bachman SP, Wilkin P, Reader T et al (2020) Extinction risk and conservation gaps for Aloe (Asphodelaceae) in the Horn of Africa. Biodivers Conserv 29:77–98. https://doi.org/10.1007/s10531-019-01870-0

Bacon E, Gannon P, Stephen S et al (2019) Aichi biodiversity target 11 in the like-minded megadiverse countries. J Nat Conserv 51:125723. https://doi.org/10.1016/j.jnc.2019.125723

Balemie K (2019) Floristic diversity of Jorgo Wato forest and ethnobotanical study of medicinal and wild edible plants in Nole Kaba district, West Wollega, Oromia region. Addis Ababa University, Ethiopia

Baranow P, Mytnik J (2009) Two new species of Polystachya Hook. (Orchidaceae) from Africa. Plant Syst Evol 281:11–16. https://doi.org/10.1007/s00606-009-0176-6

Bauters K, Goetghebeur P, Asselman P et al (2018) Molecular phylogenetic study of Scleria subgenus Hypoporum (Sclerieae, Cyperoideae, Cyperaceae) reveals several species new to science. PLoS ONE 13:e0203478. https://doi.org/10.1371/journal.pone.0203478

Belay B (2018) Vegetation distribution, land use pattern and soil seed bank composition along altitudinal gradient from Lake Tana to mount Guna, south Gondar, northern Ethiopia. Addis Ababa University

Berhanu A (2018) Vegetation ecology and conservation Status of Evergreen Afromontane Forest patches in Awi Zone of Amhara region. Addis Ababa University, Northwestern Ethiopia

BGCI (2021) State of the World's Trees. BGCI, Richmond, UK

Bisseret P, Dioli M (2005) Euphorbia bertemariae a new species from Ethiopia. Cactus Succul J 77:208–209. https://doi.org/10.2985/0007-9367(2005)77[208:EBANSF]2.0.CO;2

Bisseret P, Specks E (2006) Euphorbia gilbertiana: a new distinctive dwarf succulent euphorbia from Southern Ethiopia. Haseltonia 12. https://doi.org/10.2985/1070-0048(2006)12[15:EGANDD]2.0.CO;2

CBD (2020) Zero draft of the post-2020 global biodiversity framework. www.cbd.int/doc/c/efb0/1f84/a892b98d2982a829962b6371/wg2020-02-03-en.pdf. Accessed 11 Jul 2021

de Moura TM, Gereau R, Lewis GP (2015) Mucuna mooneyi (Leguminosae: Papilionoideae), a new species from Ethiopia. Kew Bull 70:32. https://doi.org/10.1007/s12225-015-9581-x

Demissew S, Friis I, Weber O (2021) Diversity and endemism of the flora of Ethiopia and Eritrea : state of knowledge and future perspectives. Rend Lincei Sci Fis e Nat. https://doi.org/10.1007/s12210-021-01027-8

Demissew S, Mill RR (2009) Revision of the genus Seddera (Convolvulaceae). Kew Bull 64:197. https://doi.org/10.1007/s12225-009-9112-8

Denu D, Desissa D (2013) Abundance and use of Vepris dainellii (Pichi-Serm.) Kokwaro, an Ethiopian endemic plant, in Melokoza woreda, Southern Ethiopia. Ethiop J Educ Sci 8:1–10

Friis I, Demissew S (2001) Vegetation maps of Ethiopia and Eritrea. A review of existing maps and the need for a new map for the Flora of Ethiopia and Eritrea. Biol Skr 54:399–439

Friis I, Edwards S, Kelbessa E, Demissew S (2001) Diversity and endemism in the flora of Ethiopia and Eritrea: What do the published Flora volumes tell us ? Biol Skr 54:173–193

Friis I, Gilbert MG, Paton AJ et al (2018) The Gerire Hills, a SE Ethiopian outpost of the transitional semi-evergreen bushland : vegetation, endemism and three new species, Croton elkerensis (Euphorbiaceae), Gnidia elkerensis (Thymelaeaceae), and Plectranthus spananthus (Lamiaceae). Webbia. https://doi.org/10.1080/00837792.2018.1505379

Friis I, Gilbert MG, van Breugel P et al (2017) Kalanchoe hypseloleuce (crassulaceae), a new species from eastern Ethiopia, with notes on its habitat. Kew Bull 72:30. https://doi.org/10.1007/s12225-017-9704-7

Friis I, Gilbert MG, Weber O, Demissew S (2016) Two distinctive new species Of Commicarpus (Nyctaginaceae) from gypsum outcrops in Eastern Ethiopia. Kew Bull 71:1–19. https://doi.org/10.1007/s12225-016-9648-3

Friis I, Harris T (2013) Alchornea cordifolia (Schumach. and Thonn.) Muell. Arg. (Euphorbiaceae), a disjunct Guineo-Congolian tree found in Ethiopia as dominant in riverine forest. Webbia 68:147–154. https://doi.org/10.1080/00837792.2013.857445

Friis I, Phillips SM, Gilbert MG et al (2011) New additions to the Flora of Ethiopia and Eritrea in the families Euphorbiaceae, Fabaceae (Legwninosae), Lamiaceae, Campanulaceae, Eriocaulaceae and Poaceae. Symb Bot Ups 35:107–128

Friis I, Thulin M, Adsersen H, Burger A-M (2005) Patterns of plant diversity and endemism in the Horn of Africa patterns of plant diversity and endemism in the Horn of Africa. Biol Skr 55:289–314

Gebrehiwot K, Woldu Z, Fekadu M, et al (2019) Classification and ordination of plant communities in Abune Yosef mountain range, Ethiopia. Acta Ecol Sin. https://doi.org/10.1016/j.chnaes.2019.12.001

Girmay M (2021) Ecological and ethnomedicinal study in Hirmi woodland vegetation and the surrounding districts, Tigray Regional State, Northern Ethiopia Mehari Girmay ecological and ethnomedicinal study in Hirmi Woodland vegetation and the surrounding districts. Addis Ababa University, Tigra

Gurmessa F, Soromessa T, Kelbessa E (2013) Floristic composition and community analysis of Komto Afromontane Moist Forest, East Wollega Zone, West Ethiopia. Sci Technol Arts Res Esearch J 2:58–69

Hishe M (2019) Ecological, floristic and ethnobotanical studies in and around Wejig-Mahgo-Waren massif forest patches in southern Tigray. Addis Ababa University, Ethiopia

Hoffman M, Koenig K, Bunting G, et al (2016) Biodiversity hotspots (version 2016.1). https://doi.org/10.5281/ZENODO.3261807

IUCN (2017) The IUCN Red List of threatened species. In: 2017–3. http://www.iucnredlist.org/about/summary-statistics#Tables_1_2. Accessed 28 Jun 2018

IUCN (1998) 1997 IUCN red list of threatened plants. IUCN, Gland, Switzerland and Cambridge, UK

Kassa Z (2017) Plant diversity and ethnobotanical study of medicinal and wild edible plants in Sheka zone, Southern Nations, nationalities and peoples regional state. Addis Ababa University, Ethiopia

Kebede M, Yirdaw E, Luukkanen O, Lemenih M (2013) Plant community analysis and effect of environmental factors on the diversity of woody species in the moist Afromontane forest of Wondo Genet, South Central Ethiopia. Biodivers Res Conserv 29:63–80. https://doi.org/10.2478/biorc-2013-0003

Kefalew A (2019) Ecology, carbon stock quantification and formulation of allometric models for some trees of Sele-Nono Forest, Southwest Ethiopia: implication for biodiversity conservation and climate change mitigation. Addis Ababa University

Kelbessa E (2009) Three new species of Acanthaceae from Ethiopia. Kew Bull 64:57–65. https://doi.org/10.1007/s12225-009-9097-3

Kelbessa E, Darbyshire I (2018) Six new species of Barleria L. (Acanthaceae) from Northeast Tropical Africa. Kew Bull 73:1–23. https://doi.org/10.1007/S12225-017-9725-2

Kelbessa E, Demissew S (2014) Diversity of vascular plant taxa of the flora of Ethiopia and Eritrea. Ethiop J Biol Sci 13:37–45

Maczulak A (2010) Biodiversity: Conserving endangered species. Infobase Publishing, USA

Mittermeier RA, Gil PR, Hoffmann M, et al (2004) Hotspots Revisited. CEMEX, UK

Mittermeier RA, Turner WR, Larsen FW, et al (2011) Global biodiversity conservation: the critical role of hotspots. In: Biodiversity hotspots. Springer, pp 3–22

Pankhurst R (1989) Some notes on deforestation and afforestation in Ethiopia prior to world war I. Walia 12:10–20

Paton AJ, Friis I, Demissew S (2018) A new species of Leucas, L. gypsicola (Lamiaceae), from gypsum outcrops in eastern Ethiopia. Kew Bull 73:59. https://doi.org/10.1007/S12225-018-9781-2

Peckover RG (2020) Pterodiscu an African Endemics. Cactus Succul J 92:4–11. https://doi.org/10.2985/015.092.0101

Thulin M (2021) Two new species of Commicarpus (Nyctaginaceae) from the Horn of Africa. Nord J Bot 39. https://doi.org/10.1111/njb.03115

Thulin M (2009a) New species of Caralluma and Ceropegia (Apocynaceae: Asclepiadoideae-Ceropegieae) from eastern Ethiopia. Kew Bull 64:477. https://doi.org/10.1007/s12225-009-9125-3

Thulin M (2009b) New species of Euphorbia (Euphorbiaceae) from eastern Ethiopia. Kew Bull 64:469. https://doi.org/10.1007/s12225-009-9138-y

Thulin M (2009c) Solanum pronum (Solanaceae), a new species from north-eastern tropical Africa. Kew Bull 64:491. https://doi.org/10.1007/s12225-009-9141-3

Thulin M, Goyder D, Liede-Schumann S (2008) Cibirhiza spiculata (Apocynaceae), a remarkable new species from eastern Ethiopia. Kew Bull 63:617–624. https://doi.org/10.1007/s12225-008-9060-8

Thulin M, Vollesen K (2015) Blepharis gypsophila (Acanthaceae), a new species from Ethiopia. Kew Bull 70:26. https://doi.org/10.1007/s12225-015-9578-5

Tilahun A (2018) Vegetation ecology and carbon stock of Wof-Washa Forest, North Shewa Zone, Amhara Region. Addis Ababa University, Ethiopia

Vivero J., Kelbessa E, Demissew S (2006) Progress on the red list of plants of Ethiopia and Eritrea: conservation and biogeography of endemic flowering taxa. In: Ghazanfar SA, Beentje HJ (eds) Taxonomy and ecology of African plants, their conservation and sustainable use. Royal Botanic Garden, Kew, UK, pp 761–778

Vivero JL, Kelbessa E, Demissew S (2005) The red list of endemic trees and shrubs of Ethiopia and Eritrea. Fauna and Flora International, Cambridge, UK

Wang SW, Boru BH, Njogu AW et al (2020) Floristic composition and endemism pattern of vascular plants in Ethiopia and Eritrea. J Syst Evol 58:33–42. https://doi.org/10.1111/jse.12527

Weber O, Atinafe E, Awas T, Friis I (2020) Euphorbia veneficaTrémaux ex Kotschy (Euphorbiaceae) and other shrub-like cylindrically stemmed Euphorbia with spirally arranged single spines. Bull Soc Nat Luxemb 122:57

Woldearegay M (2017) Floristic composition, structural analysis and land use/land cover change in Bore-Anferara-Wadera Forest. Addis Ababa University, Southern Ethiopia

Chapter 5
Single Large or Several Small (SLOSS)

Small is Plentiful
—Catherine Masson-Boivin, Trends in Plant Science

Abstract The question "which" and "how much" to protect as a nature reserve or protected area has been always open for debate. Due to the long-established species richness-area relationship concept, traditionally establishing a single large area as a protected area has a long history. The objective of this chapter is to synthesize the single large or several small reserves (SLOSS) hypothesis and its implication in conservation in the perspectives of Ethiopian protected areas. The protected areas in the country are *ipso facto* single large with a minimum, maximum, mean, and median area of 20.8, 46,126.7, 1889.6 and 546.8 km^2. Overall, 24 and 50% of the protected areas are greater than the mean (1889.6 km^2) and median (546.8 km^2). As the result of land cover change, habitat destruction and fragmentation, climate change, invasive species, over-exploitation, and fire, the environment in Ethiopia is highly fragmented, isolated, and covered by patchy natural vegetation. This urges the need to consider several small conservation reserves. Case studies on some small ecosystems revealed that small areas could serve as *refugia* and enhance ecological connectivity. Since the present conservation reserves might not be sufficient in the climate change era, establishing small conservation reserves could enhance connectivity and help species to find suitable habitat and withstand the ever-changing environmental conditions. If the preparedness continued similar to the current state, the protected areas would face an unprecedented crisis in the era of global warming. However, return-on-investment analyses must be undertaken prior to reserve establishment.

Keywords Climate change · Ecological connectivity · Nature reserve · Rarity · *Refugia* · Return-on-investment · SLOSS

K. G. Yaynemsa, *Plant Biodiversity Conservation in Ethiopia*,
https://doi.org/10.1007/978-3-031-20225-4_5

5.1 What is SLOSS

The question "which" and "how much" to protect as a nature reserve or protected area has been always open for debate. Although conservation at any appropriate spatial scale is possible, traditionally establishing a large area as a protected area has a long history. This is due to the long-established species richness-area relationship concept, i.e. large areas harbor a high number of species than a small area.

Conservation resources/budget is usually insufficient, particularly, in the Global South. In this era, it is challenging to achieve the protection of biodiversity with these limited conservation resources. Consequently, this leaves policy- and decision-makers at crossroads because they have to choose a relevant approach to protect biodiversity using the scarce conservation resources. The target for conservation could be either "fine-filter" in which its aim is protecting genes, species, and population or "coarse-filter" aimed at protecting communities, ecosystems, or landscape (Schwartz 1999). The "fine-filter" conservation's goal is avoiding species and population extinction in every possible way. This is supported by assessing the IUCN threat status, rarity, or other complementary approaches such as the recent work done by (Stévart et al. 2019). Since it is based on quantitative data, it is easy to evaluate the conservation progress. Hence, the measurable advantage of fine-filter conservation is that evaluation is clear; persistence is a success, whereas extinction is a failure (Schwartz 1999). The drawback of "fine-filter" is the threat status of all species is not assessed. IUCN assessment is the tip of the iceberg. For example, out of the 310, 343 described plant (including green and red algae, and mosses) species, 24, 230 (8% of the described plant species) have been evaluated by the IUCN (2017). Furthermore, all species are not taxonomically described. On the other hand, the focus of the "coarse-filter" conservation approach is the whole ecosystem/landscape and the ecological process within it. Hence, the "fine-filter" is conceptually embedded in the "coarse-filter." This is presumably a holistic approach, which protects biodiversity before it is threatened (Schwartz 1999). Lacking predetermined quantitative criteria, evaluation is an issue with this approach. Though its application is not simple, this approach is commendable.

Exclusive to the organizational level of biodiversity conservation, the single large or several small (SLOSS) dichotomy puts pressure on policy-and decision-makers. The debate is continuing in the conservation and ecology field since Diamond (1975) proposed it as a reserve design. Proponents of each hypothesis provide their evidence but the debates could not settle down yet.

5.1.1 Single Large Hypothesis

As briefly introduced in the previous section, the assumption behind single large (SL) reserve is that the larger the area the higher species richness it hosts, which is the concept of island biogeography or species richness-area relationship (MacArthur and Levins 1967). Although it is not supported by empirical data, large sites can sustain a disturbance regime supporting the full array of habitats at different stages of succession; that large sites containing many patches support a more complete array of species than a suite of isolated sites that lack interactions among patches, and that disturbance results in local extirpation of species such that recolonization is a necessity (Schwartz 1999). It is considered a holistic approach because it is believed that it preserves all levels of biodiversity such as species, ecosystems, and their interactions. Proponents of the single large reserve hypothesis argue that as reserve area increases extinction risk decreases (Schwartz 1999).

Large areas should have the potential to host new colonizers or propagule. However, how big is large? I do not think there is a clear-cut size. Several authors recommend noticeably different sizes as large. According to Noss and Cooperrider (1994), an area less than 1000 ha is considered as small and argues its contribution is minimal. However, while large intact areas can be critical for conservation, the potential value of small patches should not be ignored (Wintle et al. 2019).

5.1.2 Several Small Hypothesis

Several small (SS) reserves are favored due to three reasons (Schwartz 1999). These are:

- Prioritization of large sites discourages consideration of many critically endangered species restricted to small potential reserves,
- A second reason for not abandoning small sites is that reserve size does not necessarily predict population size nor diversity, and
- Many species are naturally restricted to small patches (rarity matters).

The several small hypotheses may not operate similarly for all taxonomic groups. Due to their mobility, mammals and birds may require large reserves than plants do. However, plants may require smaller patches. Hence, it seeks curious consideration to choose either single large or several small according to the target taxonomic group. Moreover, the financial cost and expertise required should also be rigorously examined. Some authors argue small conservation reserves are costly (Baldwin and Fouch 2018). However, in an anthropic landscape in which their natural ecosystem is extremely transformed, there are not many options to choose. Hence, the remnant small natural systems could be the only available option. According to the review by Volenec and Dobson (2019), more than 50% of the protected areas in the world are <100 ha. However, UNEP-WCMC (2019a) reported that about 15% of the protected areas are smaller than 100 ha. This shows small reserves are getting due attention globally. Although small reserves and fragments are often reported to have more generalist, synanthropic, invasive, or disturbance-adapted taxa and fewer rare, specialist species than large reserves, collections of small reserves can still hold high levels (up to 85%) of native species richness (Rösch et al. 2015). Considering the metacommunity theory, designating several small reserves is reasonably acceptable. Several small reserves could create habitat heterogeneity, which supports high taxonomic diversity.

5.2 The SLOSS Debate

The debate on the issue of SLOSS is persisting. Most of the comparisons are made based on species richness. Nevertheless, taxonomic diversity and rarity were also used as a subject of comparison (Virolainen et al. 1998). Here, SLOSS is discussed in the perspectives of species richness, taxonomic diversity, and rarity.

5.2.1 Species Richness

A study conducted in Finland showed the number of species increased in relation to the number of small mires than large mires (Virolainen et al. 1998). Irrespective of the taxonomic groups, a review article by Fahrig (2017, 2019) also revealed that species richness was higher in several small patches than in a single large patch indicating a positive impact of habitat fragmentation, which leads to several small habitats. In this review, in plants, 1 out of 30 studies reported the opposite. A study reported by Honnay et al. (1999) also reported significantly higher species richness in several small patches than single large and found no evidence of extinction risk in small patches. Similarly, Kendal et al. (2017) revealed that about 87% of all native plant species were found in small reserves, <10 ha in size. This indicated several small reserves contained a greater number of species than a few large reserves of a comparable area. The cover of native species in small reserves was no different from cover in large reserves.

5.2.2 Taxonomic Diversity

Richness and diversity are interrelated. Although it might not always be true, species richness is often linearly related to taxonomic diversity. Hence, significant reports of the positive impact of small reserves on species richness also reported a positive impact on diversity. The taxonomic diversity increased in relation to the number of small mires than a large mire (Virolainen et al. 1998). A study by Cowling and Bond (1991) on isolated reserves of 4–15 ha in South Africa Fynobs also reported that these small patches are sufficient to maintain floristic diversity for long periods.

5.2.3 *Rarity*

Species rarity is an important issue in conservation. Rare species are supposed to occur in relatively lower abundance and specific habitats. For example, they can be found in mountaintops, which is, of course, small area. Albeit the paucity in SLOSS-rarity studies, some researchers reported a significant value of small reserves in conserving rare plants. Consequently, several small reserves could play a vital role to conserve rare plants. Despite the management challenges of small reserves, they are potentially valuable to preserve rare plant species (Parker 2012). A study on wetlands in New Zealand also argued, small wetlands were critical for capturing rare and threatened species and their functional richness than their counterparts, single large reserves (Richardson et al. 2015). Similarly, Kirkpatrick and Gilfedder (1995) revealed that small patches in sub-humid Tasmania supported many rare plant species that had been eliminated from the heavily modified remainder of the landscape. Moreover, Sætersdal (1994) showed that a single large wood supports fewer regional-scale rare species than do combinations of two small woods of an equal area.

The literature still indicates environmental factors and species idiosyncrasies determine which one to use. However, many additional factors that are also important for the issue have not been well studied (Liu et al. 2019). Thus, the decision on the SLOSS debate should be based on species idiosyncratic, habitat fragmentation, habitat disturbance, habitat quality, species migration, carrying capacity, and population growth (Liu et al. 2019). Furthermore, further empirical/experimental researches are required to come to a resolution on the SLOSS debate (Fahrig 2020). The relatively convincing option could be to apply several small in fragmented landscapes (Lindenmayer et al. 2015) and single large in continuous landscapes (Liu et al. 2019). Return-on-investment (ROI) analyses to balance the ecological effectiveness of protected areas and the cost of protecting the sites should also be undertaken before its establishment (Armsworth et al. 2018).

5.3 SLOSS and Conservation in Ethiopia

Globally, biodiversity is declining at an alarming rate. This phenomenon is aggravated due to man-induced activities. Consequently, developing an approach that minimizes the human impact on biodiversity has been a prime agenda. Establishing a protected area has been an age-old mechanism applied for several decades. The question of "how much" area should be protected has been a point of discussion and is still not settled. Even though the goal of establishing a protected are might differ on the intention of what and where to preserve, cultural, political, and economic factors often determine which SLOSS approach should be established (Soule and Simberloff 1986). The notion of the IUCN which supports protecting a single large habitat is better than protecting several small patches (IUCN 1980), had influenced several countries around the world guiding conservation efforts (Fahrig 2020). Consequentially, numerous countries have been applying this principle since then.

Ethiopia is one of the tropical countries in Africa. The surface area of the country is about 1.2 million square kilometers with a wide elevation range. Even though it might vary according to the spatial resolution of digital elevation models (DEM), the elevation range of the country ranges from about 125 m below sea level to about 4533 m above sea level (m a.s.l.). Consequently, it is part of the Archipelago-like (Afromontane), Archipelago-like center of extreme floristic impoverishment (Afroalpine), and Somalia-Masai regional center of endemism (White 1983). Ethiopia is also represented in the Afrotropical ecoregion by the Ethiopian highlands (Olson and Dinerstein 2002). It also hosts two of the 36-biodiversity hotspots, namely the Eastern Afromontane and the Horn of Africa hotspots (Mittermeier et al. 2004; Hoffman et al. 2016) which are regarded as the major center of diversity and endemism for several plant species. However, a recent study showed the highest rate of population density change and lowest increase of protected area cover was found in the horn of Africa biodiversity hotspot, which is also found in Ethiopia (Cunningham and Beazley 2018). The trend is similar in the other biodiversity hotspot, Eastern Afromontane although the protected area cover showed increasing trend.

Ethiopia hosts one of the oldest protected areas in Africa, the Menagesha-Suba Forest, which was established during Emperor Zerea Yacob about 560 years ago. It is claimed as the first formally protected area in Africa (Pankhurst 1989). This is far before the publication of IUCN (IUCN 1980) that recommends single large habitat protection is more important than the counterpart, several small habitat protection. However, the status of this 95.57 km^2 National Forest Priority Area (NFPA is still proposed under the World Database on Protected Areas (WDPA). Since then Ethiopia has 138 protected areas—proposed (61, 41.20%), designated (76, 57.07%), and inscribed (1, 0.72%) (UNEP-WCMC 2019b) (Fig. 5.1). According to the definition of (UNEP-WCMC 2019b), "Proposed" means *it is in a process to gain recognition or dedication through legal or other effective means*, "designated" means *it is recognized or dedicated through legal means*, and "inscribed" means *it is only applicable for protected areas designated under the World Heritage Convention.*

The protected areas are designated differently and established for different goals. These include the National Forest Priority Area, wildlife reserve, controlled hunting area, Sanctuary, national park, UNESCO-Man and Biosphere (MAB) reserve, and World Heritage Site (Fig. 5.2). Most of the designation names are self-explanatory. Only two (2%) of the protected areas (Kafa MAB and Simien Mountains National Park) are designated as international protected areas by WDPA. Even though it is not included in the (UNEP-WCMC 2019b) document, the Lake Tana Biosphere reserve is also one of the recently documented MAB. Whereas, the remaining 98 protected areas are designated as national. These are distributed throughout the country in several ecosystems including lowlands and Afroalpine ecosystems. The area of each protected area in the country is *ipso facto* single large. The minimum, maximum, mean, and median areas of the protected areas are 20.8, 46,126.7, 1889.6 and 546.8 km^2 (UNEP-WCMC 2019b) (Table 5.1). Overall, 24 and 50% of the protected areas

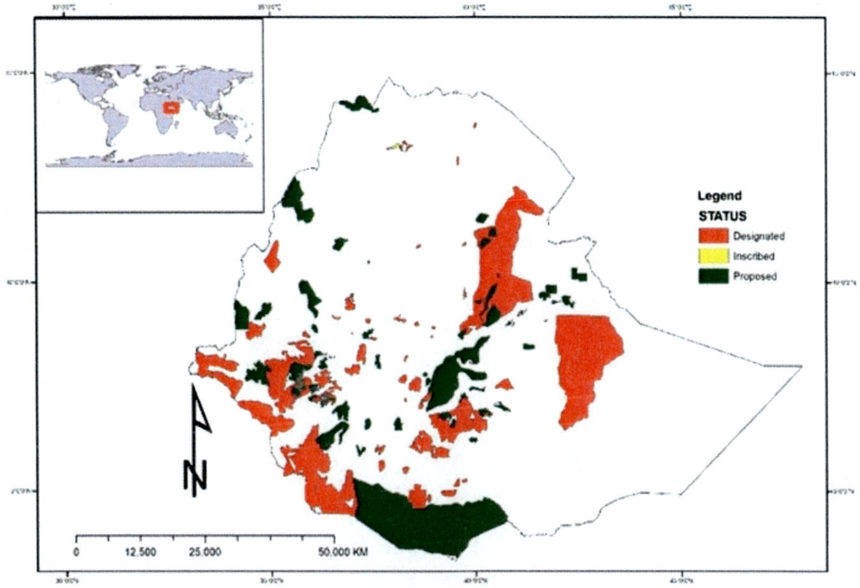

Fig. 5.1 Status of protected areas in Ethiopia. The data to produce this map were obtained from World Database on Protected Areas (UNEP-WCMC 2022)

are greater than the mean (1889.6 km^2) and median (546.8 km^2). Although it is not clearly stated whether the country is following the Single large (SL) conservation reserves or several small (SS) conservation reserves, the data revealed that Ethiopia is implementing the single large habitat approach. In order to have detailed insight about the conservation reserves including those not included in WDPA and undergo further research on the value of the small conservation reserves, I recommend the establishment of **Database of Ethiopian Conservation Reserves (DECR)**.

Even though there have been several national policies and strategic documents for biodiversity conservation in Ethiopia, neither of these records implied the SLOSS application. The National Biodiversity Strategy and Action Plan (NBSAP) prepared by the then Institute of Biodiversity Conservation, now Ethiopian Biodiversity Institute (EBI) claimed that many of the protected areas are too small and isolated to be effective; hence, Ethiopia's ecosystems are not adequately represented within the protected areas network (IBC 2005). Nevertheless, the report did not define how much area is small and what the median is or mean area of the protected areas thought to be too small. Unless it is an ill-advised conclusion, a median of 546.8 km^2 (Table 5.1) protected area cannot be considered too small. Globally, more than half of the protected areas are <100 ha in size ((Deguignet et al. 2014) though UNEP-WCMC (2019a, b) reported only 15% of the protected areas are smaller than 100 ha. Thus, the protected areas in Ethiopia are not too small, but it is true that these protected areas are too isolated, fragmented, and lacks connectivity as well. Irreplaceability, i.e., endemic species and species with a very narrow geographic range, might require

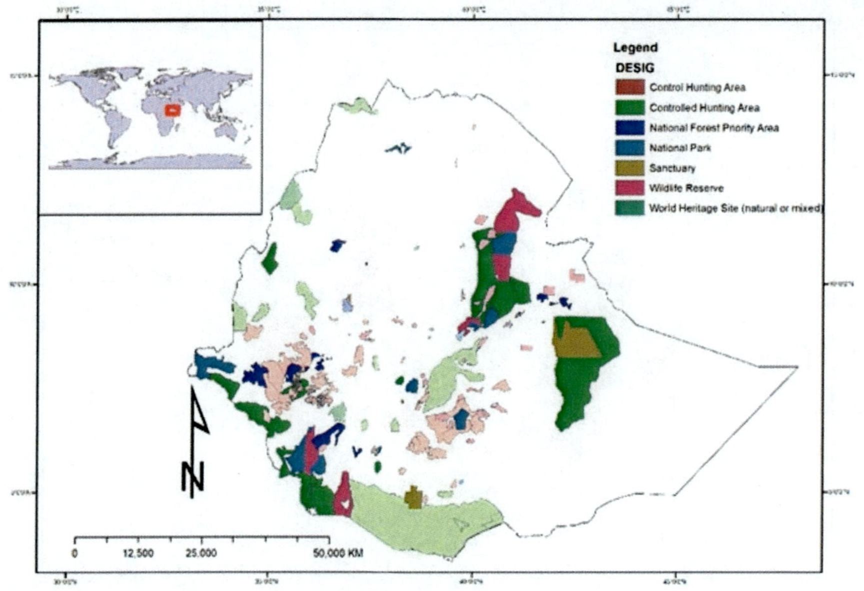

Fig. 5.2 Protected areas of Ethiopia. The data to produce this map were obtained from World Database on Protected Areas (UNEP-WCMC 2022)

Table 5.1 Descriptive statistics of the area of the protected areas in Ethiopia

	Area (km^2)
Mean	1889.6
Standard error	436.2
Median	546.8
Standard deviation	5086.7
Minimum	20.8
Maximum	46,126.7

the smallest habitat where the species is residing. In the era of climate change, range shifts should also be taken into consideration. Establishing small conservation reserves might be inevitable in the future. However, as stated by Armsworth et al. (2018) return-on-investment (ROI) analyses should be performed before its establishment.

5.3.1 *Case Studies in a Fragmented Landscape of Northern Ethiopia*

Northern Ethiopia is one of the highly degraded environments in Ethiopia. Most of its natural environment is transformed into anthropic landscapes or anthropogenic biomes. Consequently, the remnant natural vegetation is confined to either place, which is unfavorable to agriculture due to the extreme environmental conditions, or religious sites such as the Ethiopian Orthodox Tewahedo Church, which is acknowledged as a safe haven for biodiversity by some authors (Wassie and Teketay 2006). The case studies presented here are part of my Ph. D dissertation (Gebrehiwot 2020). The size of the vegetation types is approximately 70, 100 and 3700 hectares. To envisage the importance of small conservation areas in a fragmented landscape, I will roughly show the floristic composition, endemic plants recorded, threatened and rare plants in each vegetation types.

Afroalpine Grassland

The elevation above 3500 m, which is considered potentially Afroalpine belt (AA), is much converted into agricultural land. In the Abune Yosef mountain range, the Afroalpine ecosystem has been reduced over the last decades and it currently remains above 3700 m a.s.l. Agricultural practices are not uncommon at an elevation of near 3800 m a.s.l. Currently, the Afroalpine ecosystem covers only 3700 ha.

Its large numbers of geographically vicarious taxa and high endemic plant species characterize the Afroalpine belt (Hedberg 1969). The Afroalpine belt hosts plant species with life forms often different from the other vegetation types (Hedberg 1964). These life forms are giant rosette plants, tussock grasses (and sedges), acaulescent rosette plants, cushion plants, and sclerophyllous shrubs (and dwarf shrubs). Although Friis et al. (2010) reported that small trees and shrubs dominate the lower belt, the Afroalpine vegetation in the present study, which is dominantly above 3700 m a.s.l. is above the tree line. However, some tree species such as *Juniperus procera* Hochst. ex Endl. and *Hagenia abyssinica* (Bruce) J. F. Gmel. were observed around Abune Yosef monastery and homesteads at an elevation of about 3700 m a.s.l. The characteristic species of this vegetation type are mostly herbs such as the giant *Lobelia rhynchopetalum* Hemsl. and *Carex monostachya* A. Rich. The conspicuous and endemic giant *L. rhynchopetalum* was not previously reported from Welo floristic region (WU). However, it is common in the Afroalpine belt.

Ericaceous Forest

Even though both the lower and upper limits of the Ericaceous belt (EB) may vary within a small range, the ideal range of EB is between 3000 and 3200 m a.s.l. (Friis et al. 2010). Despite the elevation range differences, several authors also used different terminology such as "Ericaceous belt" (Hedberg 1969; Friis et al. 2010), "Ericaceous bushland" (White 1983), or "Ericaceous scrub" (Egziabher 1988). In the present study, this vegetation type is found in a very small remnant natural vegetation within a very steep slope. Hence, it is a disturbed and fragmented landscape which makes using the term "Ericaceous belt" difficult. Thus, in this book, the term Ericaceous forest is used.

The characteristic species of this vegetation type includes *Erica arborea* L., *Hypericum revolutum* Vahl, *Myrsine melanophloeos* (L.) R.Br. ex Sweet, *Alchemilla haumannii* Rothm, *Geranium arabicum* Forssk., *Anthemis tigreensis* J.Gay ex A.Rich., *Erigeron alpinus* L., *Haplocarpha rueppellii* K.Lewn, *Helichrysum citrispinum* Delile, *Helichrysum splendidum* Less., *Helichrysum gofense* Cufod., *Helichrysum formosissimum* Sch.Bip., *Senecio schultzii* Hochst. ex A.Rich., *Romulea fischeri* Pax, *Satureja biflora* (Buch.-Ham. ex D.Don) Briq., *Thymus schimperi* Ronniger, *Trifolium acaule* Steud. ex A.Rich., *Trifolium burchellianum* Ser., *Aloe steudneri* Schweinf., *Kniphofia foliosa* Hochst., and *Primula verticillata* Forssk (Friis et al. 2010).

Nevertheless, considerable characteristic species of the EB were found to be a characteristic species of the Afroalpine vegetation at an elevation above 3700 m a.s.l. in the present study. These are *A. tigreensis, H. rueppellii, H. citrispinum, S. schultzii, K. foliosa,* and *P. verticillata.*

On the contrary, this vegetation type shared several characteristic species with DAF/SD. Some of the shared species comprise *J. procera, Olea europaea* subsp. *cuspidata* (Wall. & G.Don) Cif., *Clutia lanceolata* Jaub. & Spach, *Discopodium penninervium* Hochst., *Maesa lanceolata* Forssk., *Rhus natalensis* Bernh., *H. abyssinica,* and *H. revolutum.*

Dry evergreen Afromontane Forest

The Dry single-dominant Afromontane Forest of the Ethiopian highlands (DAF/SD) is found at an elevation between 2500–3000 m a.s.l. Dry single-dominant Afromontane Forest of the Ethiopian highlands (DAF/SD) vegetation type is represented by characteristic species such as *J. procera, O. europaea* subsp. *cuspidata, Acokanthera schimperi* (A.DC.) Schweinf. (dominant species), *Carissa spinarum* L., *Calpurnia aurea* (Aiton) Benth., *Clausena anisata* (Willd.) Hook.f. ex Benth., *Clutia abyssinica* Jaub. & Spach, *D. penninervium, Euclea racemose* subsp. *schimperi* Murr., *Grewia ferruginea* Hochst. ex A.Rich., *M. lanceolata, M. salicifolia, Psydrax schimperianum* (A. Rich) Bridson, *Teclea nobilis* Delile., *Rhus natalensis* Bernh., *H. abyssinica, H. revolutum,* and *Urera hypselodendron* (Hochst. ex A.Rich.) Wedd (Friis et al. 2010).

However, a considerable characteristic species such as *A. schimperi, C. anisata, E. racemosa subsp. schimperi, G. ferruginea, M. salicifolia, P. schimperianum, T. nobilis, H. abyssinica,* and *H. revolutum* are missing from the present study. Furthermore, *M. salicifolia, H. abyssinica,* and *H. revolutum* were still found in the Ericaceous Forest.

These vegetation types showed a significant difference in terms of species richness, diversity, and endemic species richness. The vascular plants' rarity and the number of threatened plants also showed differences. As shown in Fig. 5.3, vascular species richness and the threatened plants are significantly higher in the relatively small area vegetation types though the endemic plant richness is higher in the Afroalpine belt with a relatively larger area.

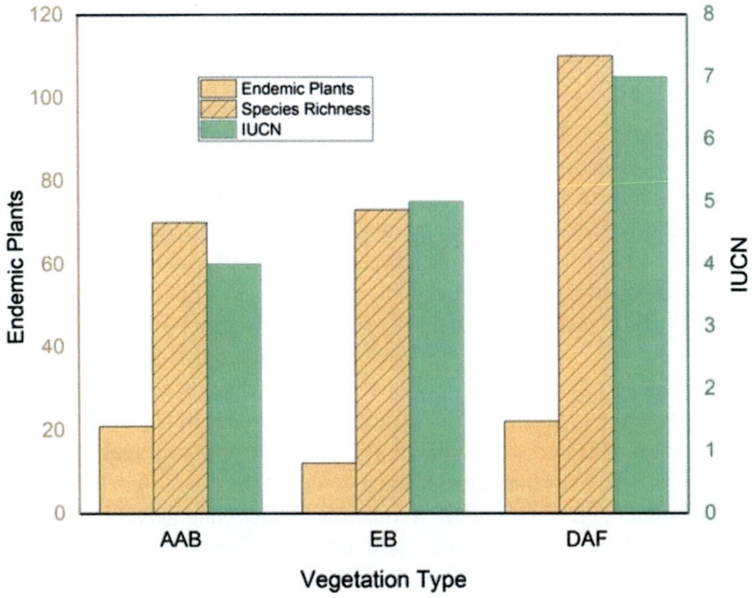

Fig. 5.3 Species richness, endemic richness, and threatened plants of the vegetation type

However, neither of these vegetation types are recognized as protected areas though the Afroalpine belt and the Dry single-dominant Afromontane forest are managed by the community and Ethiopian Orthodox Tewahedo Church, respectively (Gebrehiwot et al. 2019a, b). The Ericaceous Forest is, however, managed privately without proper conservation guidelines. Therefore, the endemic and threatened vascular plants in these fragmented landscapes could be predisposed to extinction since they are not formally acknowledged as protected areas at any level of administration.

The protection of only large and connected landscapes could miss such important vegetation and biodiversity within it. Therefore, governments and conservation organizations should consider such small areas even smaller than this area as small conservation reserves (Gebrehiwot 2020). Small conservation reserves also serve as a typical means to create habitat connectivity in fragmented landscapes. Albeit the need for return-on-investment analysis, the importance of small natural environments in biodiversity conservation should not be ignored. A shared management approach between the national and regional administration levels might be followed if the conservation cost is limited.

5.4 Threats to Biodiversity in Protected Areas in Ethiopia: SLOSS Perspective

The primary goal of protected areas is to conserve biological diversity and enhance ecosystem services and functions. Hence, nations signatory to the convention of biodiversity has agreed to increase the protected area networks coverage to 17% by 2020 in their country according to the Target 11 of the Aichi agreement (CBD 2011). This indeed recognizes the sustainable utilization of natural resources. Ethiopia is one of the few CBD signatory nations that achieved the Target 11 of the Aichi agreement (Bacon et al. 2019). It is obvious that the conservation Target 11 of the Aichi is in some sort political and achieving the target does not guarantee the desired conservation objective is achieved. For instance, a global study on changes in human population density and protected areas in terrestrial global biodiversity hotspots of two decades revealed that population density in the biodiversity hotspots averagely increased by about 36%, which exceeds the global average (Cunningham and Beazley 2018). This study also showed the highest rate of population density change and lowest increase of protected area cover was found in the horn of Africa biodiversity hotspot, which is also found in Ethiopia. The trend is similar in the other biodiversity hotspot, Eastern Afromontane, although the protected area cover showed a slight improvement. This creates doubt about the achievement of Target 11 of the Aichi, while the population density in the protected areas is increasing. It is natural to raise question "is only the area cover achieved or biological diversity, ecosystem service and functions are also improved?".

However, the post-2020 target 3 proposes that "ensure that at least 30% of the globally land areas and of sea areas, especially areas of particular importance for biodiversity and its contribution to people are conserved effectively and equitable managed, ecologically representative and well-connected systems of protected areas and other effective area-based conservation measures (OECMs) and integrated into the wider landscape and seascape" (CBD 2021).

The nation's biodiversity conservation priorities are found in the five strategic goals (EBI 2015). These are:

- Address the underlying causes of biodiversity loss by mainstreaming biodiversity across government and society;
- Reduce the direct pressures on biodiversity and promote sustainable use;
- Improve the status of biodiversity by safeguarding ecosystem, species, and genetic diversity;
- Enhance the benefits to all from biodiversity and ecosystem services; and
- Enhance implementation through participatory planning, knowledge management, and capacity building

These strategic goals are the umbrella for the 20 Aichi targets. The National Biodiversity Strategies and Action Plans (NBSAP) document was prepared in 2015 for five years. This is the final year of the NBSAP strategic goal. Is there any assessment of the NBSAP strategic goals yet? Which goals are achieved and which are not? Does the direct pressure on biodiversity reduced? Even though answering these questions needs a formal assessment of the NBSAP achievements, I will imply the persisting major threats to biodiversity in protected areas in Ethiopia. In addition, I will try to strengthen the value of small conservation reserves, as I did in the previous sections, under these pervasive threats. Emphasis will be given to protected areas and their environs.

5.4.1 Habitat Destruction and Habitat Fragmentation

Most of the protected areas in Ethiopia have been established in the 1960 and 1970s to mitigate biodiversity loss and better ensure ecosystem functions and services. The contribution of these protected areas in biodiversity conservation is undeniable. However, these protected areas are still facing several pressures, which threatened their existence. As discussed in Chap. 1, habitat destruction (conversion) and habitat fragmentation are a global phenomenon that threatens biodiversity. *Ipso facto*, Ethiopia could not escape from these threats too. The conversion of natural vegetation such as forests, woodlands, and grasslands into agricultural land has a long history in Ethiopia. The literature claimed that, in Ethiopia, the Ox-plow, which is the main tool even today, started 500 A.D (McCann 2017). Other authors extend the development of Ox-plow agriculture to 500–1000 BC and claim man-influenced vegetation, particularly, in north-central Ethiopia for more than 2000 years (Teketay et al. 2003). This is sometimes supported by palynological studies. Agricultural expansion at the expense of natural vegetation has caused a devastating impact on biodiversity including in protected area networks.

A study in Bale Mountains National Park, which is one of the largest parks in Ethiopia, reported about −7.8% (−72.3 km^2 natural forest cover loss in 23 years from 1987 to 2015), which was attributed to the expansion of small-scale subsistence agriculture (Young et al. 2020). Similarly, Arafaine and Asefa (2019) reported a decline of 39.8% (862.3 km^2) of woodland cover in Kafta-Sheraro National Park just in 12 years (2003–2015) only. A significant loss of forest cover both open and dense forest was also reported by Negassa et al. (2020) in Komto national forest priority area. The loss is attributed to agricultural expansion and conversion to grassland. The national park located in the rift valley, Nech Sar, also showed conversion of woody vegetation to grassland (5%) and subsequently grassland-to-open-overgrazed land (28.26%) (Deribew 2019). In the Bale eco-region, landscape matrix analysis revealed increased fragmentation and agricultural land expansion at the expense of dense forest and woodland (Mezgebu and Workineh 2017). Similar trends were being reported elsewhere in Ethiopia. This partly raises a question about the effectiveness of the protected areas in the country. Regardless of the Aichi Target 11 achievement, habitat destruction/conversion and fragmentation lead to the formation of several smaller patches, which are sparsely connected. This implies the need to revise the conservation approach. I genuinely recommend consideration of several small patches as conservation reserves.

5.4.2 Over-Exploitation

Unsustainable utilization of natural resources such as exploitation of indigenous tree species for timber, over-harvesting of woody species for fuelwood, and uncontrolled collection of herbal medicine are primary threats for plant diversity and the ecosystem in protected areas in Ethiopia. Reduction of natural forest has come about both through commercial exploitation and by clearance for agriculture, and surviving forests everywhere are increasingly subjected to pressure for charcoal and firewood from the rapidly expanding human population, as well as for cultivation (Atkins and Edwards 2009). Extensive exploitation of Menagesha Suba and Jibat forests for timber production, firewood, and construction purposes was reported (Berhanu 2017). In addition to the indigenous tree species mentioned in Chap. 1, species such as *Rhus glutinosa* subsp. *glutinosa* Hochst. ex A.Rich. had been excessively exploited for its high-quality wood for traditional house construction. In northern Ethiopia, over-exploitation together with fire have degraded the indigenous *Juniperus procera–Olea eurpaea*-dominated forest and a novel *Cadia purpurea–Opuntia ficus-indica* plant community emerged (Aynekulu et al. 2012).

A temporal woody species composition change of 16 years has been studied at Denkoro National Forest Priority Area (NFPA) (Awoke and Mewded 2019). Even though the overall species composition is improved over the time series, eight species that existed in 2001 were lost in the 2017 record. Furthermore, *Prunus africana* (Hook.f.) Kalkman, *J. procera, Apodytes dimidiata* E.Mey. ex Arn., *M. melanophloes, Ekebergia capensis* Sparrm., *Olinia rochetiana* A. Juss., *H. abyssinica,* and *O. europaea* subsp. *cuspidata* have been preferably over-exploited by the local people. Similarly, in the Simien Mountains National Park, which is designated as a World Heritage Site wood-cutting, is a common activity (Jacob et al. 2017).

5.4.3 Invasive Alien Species

Invasive alien species is a looming threat to biodiversity globally. Its impact is profound in environmentally degraded areas. Exclusive to some parts of south and southwestern, Ethiopia's natural environment is extremely degraded. As a result, most of the environment in the country created a fertile ground for invasive alien species, particularly, plant species. Numerous plant species have been recognized as invasive. As listed in Table 1.1, there are nearly 20 invasive plant species in Ethiopia. Although most of them are exotic, there are few native plants, which turned into invasive. I have observed some of the species in the desert and semi-desert scrubland at an elevation lower than 1000 m a.s.l. (e.g., *Parthenium hysterophorus* L. and *Prosopis juliflora* (Sw.) DC.) and Afromontane forests at an elevation above 2500 m a.s.l. This means the protected areas in the lowlands and highlands are prone to invasive species. Nevertheless, a few studies noticed the distribution and impact of invasive species in protected areas.

Yet the worst, some of the invasive species occurred as a plant community. In some protected areas, for instance, Aynekulu et al. (2016) reported a *Cadia purpurea-Opuntia ficus-indica* and *Pterolobium stellatum-Celtis africana* community in Hugumburda National forest priority area. Hence, getting these IAS inside the NFPA as a community is one of the worst conservation crises that can be envisaged. *Cadia purpurea* (G.Piccioli) Aiton is also one of the environmental degradation indicator species.

Awash National Park is one of the largest parks in Ethiopia, which is threatened by invasive species. Shiferaw et al. (2019) claimed the rate of *P. juliflora* invasion, which is found in the park, is estimated at 31,127 ha per year. There are also other invasive species such as *P. hysterophorus, Cryptostegia grandiflora* Roxb. ex R.Br., *Parkinsonia aculeata* L., *Senna occidentalis* (L.) Link, *Datura ferox* L., and *Xanthium strumarium* L. in Awash National Park (Yohannes et al. 2011). The highway to the neighboring country, Djibouti, together with the bare vegetation may have aggravated the occurrence of these invasive plant species in Awash National Park and its environs.

5.4.4 Climate Change

Climate change, sometimes interchangeably used with global warming, is a global phenomenon, and is the main threat to biodiversity (Reid et al. 2005). Globally, the temperature is increasing while rainfall is declining. The impact of climate change and species response to climate change is discussed in Chap. 1. The main responses include phenology shift, phenotypic plasticity, range shift, and extinction. If species are unable to withstand climate change, needless to say, their fate is extinction. Leaving the other species responses to climate change, I will discuss species range shift and protected areas in Ethiopia. I will indicate what policy- and decision-makers should consider in the future. Furthermore, the importance of small conservation reserves as *refugia* will be discussed.

Temperature and rainfall trend analysis in Ethiopia for the years 1983–2013 made by Adgeh (2017) revealed that temperature has already increased by 0.0265 to 0.1112 °C per year while rainfall was declining. Although similar projections were made on both minimum and maximum temperature, according to the predictions by IPCC (2014) rainfall is to likely increase by the end of the twenty-first century. The increasing temperature trend is unequivocally accepted globally. Consequently, species of several taxa are shifting their range. Tilman et al. (2002) and Kuussaari et al. (2009) implied that population dynamics will lag behind climatic trends and that an average of 40% of the range still occupied at the end of the twenty-first century will have become climatically unsuitable for the respective species, creating an extinction debt. Apart from the long-established arguments on the impact of climate change and how species respond, *abrupt changes in ecological systems (AES)—changes that are fast in time or fast relative to their drivers—are ubiquitous and increasing in frequency* (Turner et al. 2020). This indicates that the impact of climate change on biodiversity and ecological systems is by far complicated than we thought and seeks curious attention.

Probably, Woldu (1991) implied the impact of climate change on grasses for the first time in Ethiopia. This study implied the *C*4 plants did shift their range into higher elevation in the Danakil depression as a result of warming. This implied it could be due to global warming. It's about 30 years since this study was conducted. Despite the species distribution models (SDM), there are no other investigations regarding species (plant) range linked to climate change.

However, it is hardly observed that conservation approaches in Ethiopia considered this phenomenon. In fact, most of the protected areas were established in the 1960 and 1970s before the issue of climate change attracted attention from scientists and politicians. Thus, it is not reasonable to criticize them for not considering climate change at that time. However, lately established protected areas followed a similar trend, simply protecting a large un-fragmented natural vegetation. The global climate change impact on plant species did not influence the policy-and decision-makers to re-consider the establishment of protected area networks by considering small conservation reserves.

Several species are shifting their range toward higher altitudes and poleward (Chap. 1). As a result, transient plant communities are being formed (Schippers et al. 2021). Elevation and land surface area are often inversely correlated. Even though Ethiopia consists 50% of the elevation higher than 2000 and 80% of the elevation higher than 3000 in tropical Africa (Yalden and Largen 1992), these proportion doesn't correspond at the national level. The low altitude surface area (<2000 m a.s.l.) covers by far a large area than the high altitude surface area (>2000 m a.s.l.) (Fig. 5.4). The majority of the protected area network in Ethiopia are established at lower elevations. Therefore, it is most likely that the species from the lower elevations are to shift their range to the high elevation areas, while the high elevation species are predisposed to the climate change threats.

Hence, what could be the fate of species shifting their niche out of the protected areas as a result of climate change? Is there any conservation approach implemented that considered the climate change issue? What about the cold-adapted plants (e.g., the giant rosette plant- *L. rhynchopetalum*) in the Afroalpine belt and Ericaceous belt? Decision- and policy-makers in the era of anthropogenic-induced climate change must consider these questions. A recent study by Chala et al. (2016) predicted the impact of climate warming on habitat and genetic diversity loss in cold-adapted plants using *L. rhynchopetalum* as a representative plant species. The study revealed that only 3.4% of its habitat seems to remain suitable on average by 2080, resulting in a loss of 82% of its genetic diversity. What is even worst is that the remaining suitable habitat is more likely to be fragmented in the future. Interestingly, a study by Kidane et al. (2019) examined the potential future impacts of climate change on cold-adapted plants along an elevational gradient ranging from 2000 m a.s.l. to the highest elevation (4385 m asl) in different scenarios revealed that the endemic plants in the Afroalpine belt would be extinct. Similarly, a habitat suitability modeling on the indigenous tree species, *J. procera*, revealed that suitable habitat of the species will be decreased by

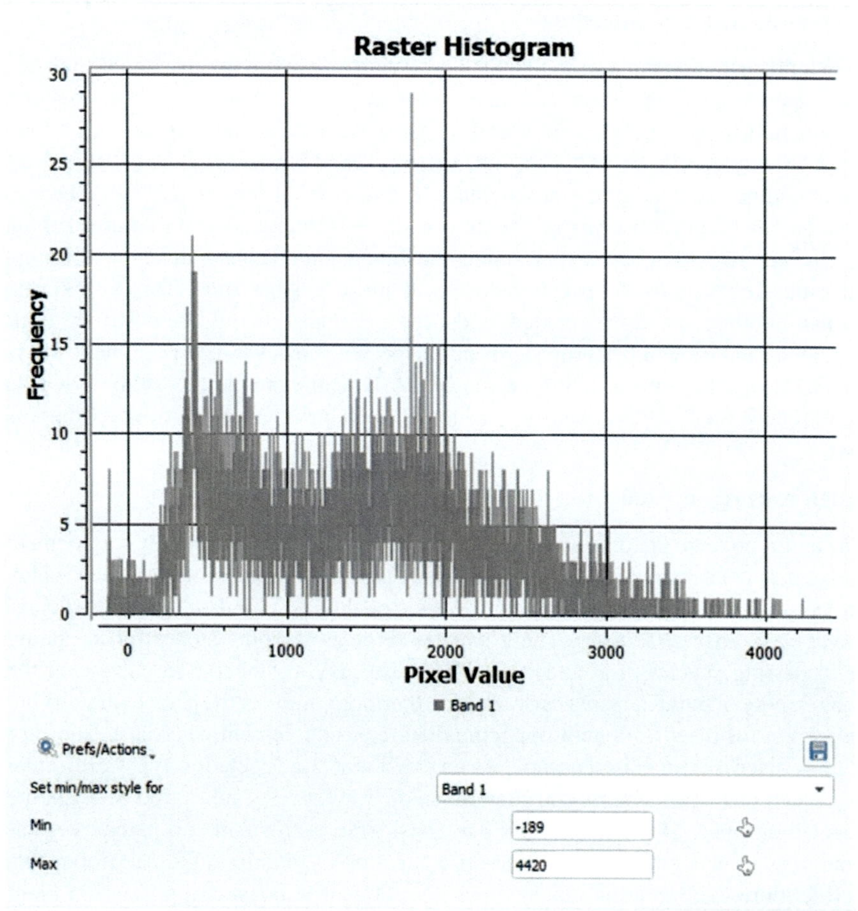

Fig. 5.4 Elevation classification of Ethiopia surface area (Digital Elevation Model). The x-axis is elevation in meters. The DEM resolution is 30 m

more than 80% and 90% by the Mid-century RCP2.6 and Mid-century RCP8.5 of the twenty-first century scenarios, respectively (Abraha et al. 1991). It is pity that the fate of the endemic species would be most likely extinction. Although the surface area of the lowlands is wider than the highlands, it often depends on the niche of the species. Thus, future conservation planning and prioritization should consider the impact of climate change on biological diversity and establishment of conservation reserves for species shifting their niche outside of the existing protected areas.

To minimize the impact of climate change on biodiversity in protected areas in general and species in particular, I firmly encourage the following considerations in the future.

Species distribution modeling

Species distribution modeling (SDM) plays a vital role in predicting suitable habitats for a species using environmental and biological or disturbance variables. Species distribution, hence, helps policy- and decision-makers in updating protected area boundaries, creating new protected area networks, and further achieving CBD targets by including the changing environmental scenarios (Vincent et al. 2019). Due to land-use and land cover change, finding a single, large, and intact natural habitat is challenging. As a result, predicting the future suitable habitat under different climatic scenarios for a species helps to identify feasible sites. This is a key in human-dominated and fragmented landscapes. Perhaps, it could be costly in terms of finance and time to perform SDM for all species. However, it is possible to study the SDM of a flagship or keystone species. A typical example is a study by Chala et al. (2016) on *L. rhynchopetalum,* which is the flagship species of the Afroalpine belt.

Small reserves as Refugia

The anthropogenic factors left the environment in Ethiopia with a highly fragmented, isolated, and patchy natural vegetation located in either religious sites or areas, which are not suitable for farming activities (Wassie et al. 2009; Chala et al. 2016; Gebrehiwot et al. 2018, 2019a, b). These patches serve as stepping stones for migrating or dispersing species in a landscape. They also play a vital role in enhancing the connectivity of protected area networks. Furthermore, some of the patches are serving as *refugia* for trees from anthropogenic disturbances (Gebrehiwot 2020). The case studies discussed in subsection 5.3.1. revealed that the vegetation types are found as a patch in a landscape. A considerable number of species are shared between the vegetation types. Therefore, giving due consideration to small and patchy vegetation types or ecosystems for conservation must be inevitable in the Anthropocene. Furthermore, conservation management should gear to considering transient plant communities. However, the return-on-investment analysis should not be ignored.

5.4.5 *Fire*

Fire is an important abiotic component that shapes an ecosystem. It plays a crucial role as a regulating factor of vegetation distribution patterns (van Breugel et al. 2016). It is obvious that there are even fire-dependent ecosystems. The tropics frequently experience both wild and human-induced fire incidents. However, the temperate region also sometimes experiences fire incidents. It's obvious that some ecosystems such as the Savannah are adapted to recurrent fire regimes and the Orthodox seeds benefit the most out of it. Although moderate fires and controlled fire regimes might have a positive influence on plant diversity, an extensive fire might lead to the extinction of rare and threatened plants and animals. Fire-sensitive ecosystems often entertain a devastating damage.

It might also lead to irrecoverable ecological harm by destroying indigenous trees, which took them hundreds of years to grow (Lemessa and Perault 2011). Moreover, a considerable number of fire-sensitive species could be easily eliminated by fire if proper management is not implemented (Tesfaye et al. 2004). Ethiopia is one of the tropical countries, which entertains the fire phenomenon almost every year though most of them are spatially small scale. The lowland protected and unprotected areas are often the target of either wild or human-induced fire. Nevertheless, recently high elevation grasslands are also becoming the target.

In Ethiopia, above 100,000 ha of forest was burnt only in the year 2000 (George and Mutch 2001). Furthermore, in their 16 years (2003–2013) study, van Breugel et al. (2016) revealed about 19% of the surface area of Ethiopia has been affected by at least one fire incident in the last 11 years, with an annual percent area burned of 6–8%. These fires affected several NFPAs in the country. Most (84%) human-induced fire incidents that happened in Bale Mountains National Park were inside the park (Belayneh et al. 2013). The threat of fire on protected areas was also reported in other parts of the country (Aynekulu et al. 2012; Siraj et al. 2017; Awoke and Mewded 2019). The degradation of indigenous plant communities by fire and other factors is the main reason for the invasive communities that occurred in that area (Aynekulu et al. 2012). Wesche et al. (2000) claimed that the patchy appearance of the Ericaceous vegetation in East Africa is due to the frequent human-induced recurrent fire incidents.

The Ethiopian government's preparedness to overcome fire incidents is, nevertheless, terrible. The fire accidents happened in Bale Mountains National Park a few years ago and Simien Mountains National Park (World Heritage Site) in 2019 is a typical example. If the preparedness continued similar to the current state, the protected areas would face an unprecedented crisis in the era of global warming.

References

Abraha S, Beyene A, Dubale T et al (1991) What factors shape girls' school peformance? evidence from Ethiopia. Int J Educ Dev 11:107–118. https://doi.org/10.1016/0738-0593(91)90036-8

Adgeh TM (2017) Characterization of present day climate trend over Ethiopia for impact study. In: Filho WL, Simane B, Kalangu J et al (eds) Climate change adaptation in Africa: fostering resilience and capacity to adapt. Springer International Publishing AG, Gewerbestrasse, Switzerland

Arafaine Z, Asefa A (2019) Dynamics of land use and land cover in the Kafta-Sheraro national park, NW Ethiopia: patterns, causes and management implications. Momona Ethiop J Sci 11:239–257

Armsworth PR, Jackson HB, Cho S et al (2018) Is conservation right to go big ? Protected Area Size and Conservation. 225:229–236. https://doi.org/10.1016/j.biocon.2018.07.005

Atkins J, Edwards S (2009) The vegetation of Ethiopia and Eritrea. In: Birds of Ethiopia and Eritrea: an atlas of distribution. Christopher Helm, London

Awoke H, Mewded B (2019) Changes in woody species composition and structure of Denkoro dry evergreen Afromontane forest over 16 years (2001–2017), South Wollo, Ethiopia. For Ecol Manage 441:71–79

Aynekulu E, Aerts R, Denich M, et al (2016) Plant diversity and regeneration in a disturbed isolated dry Afromontane forest in northern Ethiopia. Folia Geobot. https://doi.org/10.1007/s12224-016-9247-yPlant

Aynekulu E, Aerts R, Moonen P et al (2012) Altitudinal variation and conservation priorities of vegetation along the Great Rift Valley escarpment, Northern Ethiopia. Biodivers Conserv 21:2691–2707

Bacon E, Gannon P, Stephen S, et al (2019) Aichi Biodiversity Target 11 in the like-minded megadiverse countries. J Nat Conserv 51:125723. https://doi.org/10.1016/j.jnc.2019.125723

Baldwin RF, Fouch NT (2018) Understanding the biodiversity contributions of small protected areas presents many challenges. Land 7:123. https://doi.org/10.3390/land7040123

Belayneh A, Yohannes T, Worku A (2013) Recurrent and extensive forest fire incidence in the Bale Mountains National Park (BMNP), Ethiopia: extent, cause and consequences. Int J Environ Sci 2:29–39

Berhanu A (2017) Vegetation ecology and conservation status of evergreen Afromontane forest patches in Awi zone of Amhara Region, Northwestern Ethiopia. Addis Ababa University, Ethiopia

CBD (2011) COP decision X/2: strategic plan for biodiversity 2011–2020

CBD (2021) First draft of the post-2020 global biodiversity framework. In: The convention on biological diversity. https://www.cbd.int/doc/c/914a/eca3/24ad42235033f031badf61b1/wg2020-03-03-en.pdf. Accessed 25 Dec 2021

Chala D, Brochmann C, Psomas A et al (2016) Good-bye to tropical alpine plant giants under warmer climates? loss of range and genetic diversity in Lobelia rhynchopetalum. Ecol Evol 6:8931–8941. https://doi.org/10.1002/ece3.2603

Cowling RM, Bond WJ (1991) How small can reserves be ? an empirical approach in Cape Fynbos, South Africa. Biol Conserv 58:243–256

Cunningham C, Beazley KF (2018) Changes in human population density and protected areas in terrestrial global biodiversity hotspots, 1995–2015. Land 7:136. https://doi.org/10.3390/land7040136

Deguignet M, Juffe-Bignoli D, Harrison J et al (2014) 2014 United Nations list of protected areas: Technical report. Cambridge, UK

Deribew KT (2019) Spatially explicit statistical modeling of random and systematic land cover transitions in the main grassland plain of Nech Sar National Park, Ethiopia. Ecol Process 8:46. https://doi.org/10.1186/s13717-019-0199-z

Diamond JM (1975) The island dilemma: lessons of modern biogeographic studies for the design of natural reserves. Biol Conserv 7:129–146

EBI (2015) Ethiopia's national biodiversyty starategy and action plan 2015–2020. Addis Ababa, Ethiopia

Egziabher TBG (1988) Vegetation and environment of the mountains of Ethiopia: implications for utilization and conservation. Mt Res & Dev 8:211–216

Fahrig L (2017) Ecological responses to habitat fragmentation per Se. Annu Rev of Ecol Evol Syst 48:1–23. https://doi.org/10.1146/annurev-ecolsys-110316-022612

Fahrig L (2019) Habitat fragmentation: a long and tangled tale. 33–41. https://doi.org/10.1111/geb.12839

Fahrig L (2020) Why do several small patches hold more species than few large patches ? 1–14. https://doi.org/10.1111/geb.13059

Friis I, Demissew S, Breugel P van (2010) Atlas of the potential vegetation of Ethiopia. The Royal Danish Academy of Sciences and Letters, Denmark

Gebrehiwot K (2020) Vegetation ecology and dynamics of land cover change of Abune Yosef mountain range, Lasta district, Amhara regional state, northern Ethiopia. Addis Ababa University

Gebrehiwot K, Demissew S, Woldu Z et al (2019a) Elevational changes in vascular plants richness, diversity, and distribution pattern in Abune Yosef mountain range, Northern Ethiopia. Plant Divers 41:220–228. https://doi.org/10.1016/j.pld.2019.06.005

Gebrehiwot K, Demissew S, Woldu Z et al (2019b) Elevational changes in vascular plants richness, diversity, and distribution pattern in Abune Yosef mountain range, Northern Ethiopia. Plant Divers 41:220–228. https://doi.org/10.1016/j.pld.2019b.06.005

Gebrehiwot K, Desalegn T, Woldu Z et al (2018) Soil organic carbon stock in Abune Yosef Afroalpine and sub-afroalpine vegetation, Northern Ethiopia. Ecol Process 7:6. https://doi.org/10.1186/s13717-018-0117-9

George CW, Mutch RW (2001) Ethiopia: strengthening forest fire management. Italy, Rome

Hedberg O (1969) Evolution and speciation in a tropical high mountain flora. Biol J Linn Soc 1:135–148

Hedberg O (1964) Features of Afroalpine plant ecology. Almqvist & Wiksells Boktryckeri AB, Uppsala

Hoffman M, Koenig K, Bunting G et al (2016) Biodiversity hotspots (version 2016.1). https://doi.org/10.5281/ZENODO.3261807

Honnay O, Hermy M, Coppin P (1999) Effects of area, age and diversity of forest patches in Belgium on plant species richness, and implications for conservation and reforestation. Biology 87:73–84

IBC (2005) In: National biodiversity strategy and action plan. Addis Ababa, Ethiopia

IPCC (2014) Climate change 2014 impacts, adaptation, and vulnerability part B: regional aspects regional aspects. contribution of working group II to the fifth assessment report of the intergovernmental panel on climate change. Cambridge University Press, Cambridge, United Kingdom

IUCN (2017) The IUCN red list of threatened species. I2017-3. http://www.iucnredlist.org/about/summary-statistics#Tables_1_2. Accessed 28 Jun 2018

IUCN (1980) World conservation strategy. IUCN, Geneva, Switzerland

Jacob M, Frankl A, Hurni H et al (2017) Land cover dynamics in the Simien Mountains (Ethiopia), half a century after establishment of the National Park. Reg Environ Chang 17:777–787

Kendal D, Zeeman BJ, Ikin K et al (2017) The importance of small urban reserves for plant conservation. Biol Conserv 213:146–153. https://doi.org/10.1016/j.biocon.2017.07.007

Kidane YO, Steinbauer MJ, Beierkuhnlein C (2019) Dead end for endemic plant species? a biodiversity hotspot under pressure. Glob Ecol Conserv 19:e00670. https://doi.org/10.1016/j.gecco.2019.e00670

Kirkpatrick JB, Gilfedder L (1995) Maintaining integrity compared with maintaining rare and threatened taxa in Remnant Bushland in subhumid Tasmania. Biol Conserv 74:1–8

Kuussaari M, Bommarco R, Heikkinen RK et al (2009) Extinction debt: a challenge for biodiversity conservation. Trends Ecol Evol 24:564–571. https://doi.org/10.1016/j.tree.2009.04.011

Lemessa D, Perault M (2011) Ethiopia: forest fires in Ethiopia: reflections on socioeconomic and environmental effects of the fires in 2000. Ethiopia

Lindenmayer DB, Wood J, Mcburney L et al (2015) Forest ecology and management single large versus several small: the SLOSS debate in the context of bird responses to a variable retention logging experiment. For Ecol Manage 339:1–10. https://doi.org/10.1016/j.foreco.2014.11.027

Liu H, Li W, Lv G (2019) How nonrandom habitat loss affects nature reserve planning strategies. Ecol Modell 397:39–46. https://doi.org/10.1016/j.ecolmodel.2018.12.014

MacArthur R, Levins R (1967) The limiting similarity, convergence, and divergence of coexisting species. Am Nat 101:377–385. https://doi.org/10.1177/002114009105700104

McCann JC (2017) The history of agriculture in Ethiopia. Oxford Res Encycl African Hist 1–14

Mezgebu A, Workineh G (2017) Changes and drivers of afro-alpine forest ecosystem: future trajectories and management strategies in Bale eco-region, Ethiopia. Ecol Process 6:42

Mittermeier RA, Gil PR, Hoffmann M et al (2004) Hotspots Revisited. CEMEX, UK

Negassa MD, Mallie DT, Gemeda DO (2020) Forest cover change detection using Geographic Information Systems and remote sensing techniques: a spatio-temporal study on Komto Protected forest priority area, East Wollega Zone, Ethiopia. Environ Syst Res 9:1

Noss RF, Cooperrider A (1994) Saving nature's legacy: protecting and restoring biodiversity. Island Press

Olson DM, Dinerstein E (2002) The global 200: priority ecoregions for global conservation. Ann Missouri Bot Gard 89:199–224

Pankhurst R (1989) Some notes on deforestation and afforestation in Ethiopia prior to world war I. Walia 12:10–20

Parker S (2012) Small reserves can successfully preserve rare plants despite management challenges. Nat Areas J 32:403–411. https://doi.org/10.3375/043.032.0409

Reid WV, Mooney HA, Cropper A et al (2005) In: Ecosystems and human well-being-synthesis: a report of the millennium ecosystem assessment. Island Press

Richardson SJ, Clayton R, Rance BD et al (2015) Small wetlands are critical for safeguarding rare and threatened plant species. Appl Veg Sci 18:230–241. https://doi.org/10.1111/avsc.12144

Rösch V, Tscharntke T, Scherber C, Batáry P (2015) Biodiversity conservation across taxa and landscapes requires many small as well as single large habitat fragments. Oecologia.https://doi.org/10.1007/s00442-015-3315-5

Sætersdal M (1994) Rarity and species/area relationships of vascular plants in deciduous woods, western Norway-applications to nature reserve selection. Ecography (cop) 17:23–38

Schippers P, Leyequien E, Jana A (2021) Biodiversity conservation in climate change driven transient communities. Biodivers Conserv 2885–2906. https://doi.org/10.1007/s10531-021-02241-4

Schwartz MW (1999) Choosing the appropriate scale of reserves for conservation. Annu Rev Ecol Syst 30:83–108

Shiferaw H, Bewket W, Alamirew T et al (2019) Implications of land use/land cover dynamics and Prosopis invasion on ecosystem service values in Afar Region, Ethiopia. Sci Total Environ Science 67:354–436. https://doi.org/10.1016/j.scitotenv.2019.04.220

Siraj M, Zhang K, Sebsebe D, Zerihun W (2017) Floristic composition and plant community types in Maze National Park, southwest Ethiopia. Appl Ecol Environ Res 15:245–262. https://doi.org/10.15666/aeer/1501_245262

Soule M, Simberloff D (1986) What do genetics and ecology tell us about the design of nature reserves? Biol Conserv 35:19–40

Stévart T, Dauby G, II PPL et al (2019) A third of the tropical African flora is potentially threatened with extinction. Sci Adv 5:eaax9444

Teketay D, Fetene M, Abate A (2003) State of the environment in Ethiopia: past, present and future prospects. In: Environment and environmental change in Ethiopia. Semen Hotel, Addis Abeba, Ethiopia, pp 1–36

Tesfaye G, Teketay D, Assefa Y, Fetene M (2004) The impact of fire on the soil seed bank and regeneration of Harenna forest, Southeastern Ethiopia. Mt Res Dev 24:354–361

Tilman D, May RM, Lehman CL, Nowak MA (2002) Habitat destruction and the extinction debt. Nature 371:6566

Turner MG, Calder WJ, Cumming GS et al (2020) Climate change, ecosystems and abrupt change: science priorities. Philos Trans R Soc B 375:20190105

UNEP-WCMC (2019a) The world database on protected areas: Technical Report. Cambridge, UK

UNEP-WCMC (2019b) User manual for the world database on protected areas and world database on other effective area-based conservation measures: 1.6. UNEP-WCMC, Cambridge, UK

UNEP-WCMC (2022) Protected Area Profile for Ethiopia from the World Database on Protected Areas. https://www.protectedplanet.net/

van Breugel P, Van FI, Demissew S et al (2016) Current and future fire regimes and their influence on natural vegetation in Ethiopia. Ecosystems 19:369–386. https://doi.org/10.1007/s10021-015-9938-x

Vincent C, Fernandes RF, Cardoso AR, et al (2019) Climate and land-use changes reshuffle politically-weighted priority areas of mountain biodiversity. Glob Ecol Conserv 17:e00589. https://doi.org/10.1016/j.gecco.2019.e00589

Virolainen KM, Suomi T, Suhonen J, Kuitunen M (1998) Conservation of vascular plants in single large and several small mires: species richness, rarity and taxonomic diversity. J Appl Ecol 35:700–707

Volenec ZM, Dobson AP (2019) Conservation value of small reserves. Conserv Biol 0:1–14. https://doi.org/10.1111/cobi.13308

Wassie A, Sterck FJ, Teketay D, Bongers F (2009) Tree regeneration in Church forests of Ethiopia: Effects of microsites and management. Biotropica 41:110–119. https://doi.org/10.1111/j.1744-7429.2008.00449.x

Wassie A, Teketay D (2006) Soil seed banks in church forests of northern Ethiopia : Implications for the conservation of woody plants. Flora 201:32–43. https://doi.org/10.1016/j.flora.2005.04.002

Wesche K, Miehe G, Kaeppeli M (2000) The significance of fire for Afroalpine Ericaceous vegetation. Mt Res Dev 20(20):340–347

White F (1983) The vegetation of Africa. In: A descriptive memoir to accompany the UNESCO/AETFAT/UNESCO vegetation map of Africa. UNESCO, Paris, France

Wintle BA, Kujala H, Whitehead A et al (2019) Global synthesis of conservation studies reveals the importance of small habitat patches for biodiversity. Proc Natl Acad Sci United States Am 116:909–914. https://doi.org/10.1073/pnas.1813051115

Woldu Z (1991) Implications of photosynthetic pathways of some grass species in Ethiopia to land use. Coenoses 6:127–130

Yalden D, Largen M (1992) The endemic mammals of Ethiopia. Mammal Rev 22:115–150

Yohannes T, Awas T, Demissew S (2011) Survey and documentation of the potential and actual invasive alien plant species and other biological threats to biodiversity in Awash National Park, Ethiopia. Manag Biol Invasions 2

Young NE, Evangelista PH, Mengitsu T, Leisz S (2020) Twenty-three years of forest cover change in protected areas under different governance strategies : a case study from Ethiopia's southern highlands. Land use policy 91:104426. https://doi.org/10.1016/j.landusepol.2019.104426

Chapter 6
Pollination and Dispersal in Fragmented Landscape

Pollination ecology research is a neglected topic in the Ethiopian academia.

Abstract Pollination and dispersal are some of the important ecological functions of an ecosystem. Above 90% of all flowering plant species are pollinated by animals. Pollinators are drivers of plant distribution and assemblage into plant communities. However, these, often mutualistic interactions, are threatened in the Anthropocene. Plant–pollinator interaction in a fragmented landscape could cause community/ecosystem instability by decreasing productivity. Conserving remnant patches as conservation reserves would attract several pollinators, which in turn promotes pollination. In the long term, it would maintain and increase plant diversity of an ecosystem. Gene flow within and among plant communities is largely shaped by seed dispersal. This further enhances connectivity between plant populations. Threats such as habitat fragmentation could lead to the loss of biotic dispersers such as mammals and birds, which could lead to the loss of plant species and functional diversity by reducing seed dispersal and recruitment. Dispersers' loss could have direct and indirect influences on plant population and communities. Establishing ecological networks to improve the success of seed dispersal through a fragmented landscape could increase ecological connectivity. Several small reserves (SS), trees scattered in farmlands, Church forests, and home gardens could be considered as seedscape for plant dispersal in a human-dominated landscape. In order to increase the probability of seeds reaching suitable habitats and increase species richness and diversity, the protection of smaller patches as small reserves could serve as habitat connectivity. Nevertheless, great care must be applied to avoid the patches not to serve as a dead-end for the dispersers and limit seed dispersal.

Keywords Biotic dispersers · Ecological connectivity · Habitat fragmentation · Plant–pollinator interactions · Seed dispersal · Seedscape

K. G. Yaynemsa, *Plant Biodiversity Conservation in Ethiopia*,
https://doi.org/10.1007/978-3-031-20225-4_6

6.1 Pollination Ecology

Ecosystems entertain very complex interactions. These include the interactions between and within biodiversity and the ecological complexes where they are living. Pollination and dispersal are some of the important ecological functions of an ecosystem. Pollination is a mechanism in which a pollen grain is transported from an anther (male reproductive organ) to a stigma (female reproductive organ) either via pollination agents such as winds, water, and animals in the case of cross-pollination or species idiosyncratic in the case of self-pollination.

However, most angiosperms are animal pollinated. According to Ollerton et al. (2011), nearly 90% of the flowering plants are pollinated by invertebrates and vertebrates. Bawa (1990) also estimated approximately 98–99% of all flowering plant species in the tropical lowland rain forest are pollinated by animals. Pollination, often a mutualistic interaction, is one of the most important ecosystem services. Animals and winds pollinate most terrestrial flowering plants. Thus, pollinators are simultaneously vital to supporting both natural ecosystems and human food security, which is a unique position for such a diverse group of organisms (Ollerton 2017).

However, these interactions are threatened due to dominantly anthropogenic activities. The major threats to plant–pollinator interactions are more or less similar to the threats discussed in Chap. 1. Land-use change, such as habitat loss and fragmentation, and climate change are some determinant threats to plant–pollinator interactions. However, ecological restoration practices remain challenging because of uncertainty about how landscape quality influences pollinators' populations across their complex life cycle (Lozier 2017). Numerous factors such as climate change have caused a significant damage on pollinators.

Phenological mismatch of plant–pollinator due to climate change (Settele et al. 2016) has been reported globally. Species range shift of either the plant or the pollinator also causes decoupling of the plant–pollinator interaction spatially. As stated in previous sections, particularly, Sect. 3.4.4, designing a protected area and connectivity must consider species range shift. In addition, pollinators should also be considered in every restoration and conservation effort. Even though the decreasing trend of species richness from the equator toward the poles is recognized since the nineteenth century (Hillebrand 2004), several species are ranging their shift toward the poles as a result of global warming (Lenoir et al. 2008; Chen et al. 2011).

In the Global South in general and Ethiopia in particular, research on the range and phenology shift of plants and pollinators is scanty. However, in Europe and North America, this issue is relatively well researched. For example, Ollerton (2017) reviewed pollinator diversity distribution, ecological function, and conservation, in which most of the articles reviewed are from the Global North. Furthermore, other researchers also investigated climate change impacts on the range and phenology shifts (Nakazawa and Doi 2012; Kerr et al. 2015; Rasmont et al. 2015; CaraDonna et al. 2020) even though bumblebees are the most frequently studied pollinators. The other problem in associating pollination with conservation is that there is a paucity of empirical research on the pollination of wild plants both on global and regional scales.

Most of the pollination studies are related to agriculture and agricultural productivity. These studies are also based on honeybees (*Apis mellifera*) and bumblebee species. Consequently, there is a gap in the plant–pollinator studies on natural ecosystems such as forests and grasslands. Understanding this aspect is crucial to improve future conservation planning.

Pollinators are drivers of plant distribution and assemblage into plant communities (Pellissier et al. 2012). The debate about what shapes plant community has long been established since the beginning of the twentieth century (Clements 1916; Gleason 1926) and is not settled down yet. Most of the studies on plant communities focused on topographic variables, climatic variables, and edaphic (soil physico-chemical) properties. In addition, biotic interactions such as competition and facilitation were also received considerable attention. Plant–pollinator networks can potentially modify the population dynamics and the occupied range of a plant species (Pellissier et al. 2012). However, the contribution of pollination networks as a driver of plant distribution and assemblage of plant communities has received little attention. Thus, there is a strong need to characterize plant–pollinator interactions at large spatial scales and especially with respect to dynamic communities, whose compositions and patterns of relative species abundance vary in time and space (Pellissier et al. 2012). Pollinator diversity and abundance play a crucial role in shaping plant communities.

However, studies revealed that pollinators' diversity is declining at an alarming though most of the researches are on crop pollinators. In the Anthropocene, the globe is warming. This is causing changes in species (plant and/or pollinator) fundamental and realized environmental niches. As described in the previous chapters, plants are shifting their ranges and phenology. Consequently, this causes plant–pollinator mismatches which leads to plant and pollinators diversity decline. Even though it is beyond the scope of this book, it might be also an underlying driver for speciation. The most species at risk could be plants that require specialized pollinators while the generalized ones are least affected. This is due to specialized plants will be more prone to pollen limitation because they are less likely to interact with any available pollinator (Knight et al. 2005). Some small fragments could probably support unique plant–pollinator interactions that are not found in large fragments across a landscape (Librán-Embid et al. 2021). This implies that the impact of low abundance and diversity of pollinators do not only influence individual plant species but also extend to the unique plant–pollinator interactions in the plant communities.

Plant community structure and composition could also be affected by a reduction in the diversity of pollinators' functional groups (Fontaine et al. 2006). Pollinator availability could also drive the floral diversity along an elevation gradient (Pellissier et al. 2012). For example, a study on the alpine landscape in the western Swiss Alps on the relationship of pollinators and plant distribution revealed that insect pollinators are highly correlated to elevation (Pellissier et al. 2010). Plant–pollinator networks are also the main components of ecosystem stability. If the pollination type is, however, obligate mutualism, it could lead to ecosystem instability. This could happen for two reasons. (1) Co-extinction: The loss of one of the interacting partners in species-specific interactions may lead to the extinction of the other or (2) indirect effect: For instance, the loss of a fig species and its pollinating wasp species may also lead to a loss of the non-pollinating wasp species that parasitize the pollinating species (Bawa 1990).

Plant–pollinator interaction in a fragmented landscape could cause community/ecosystem instability by decreasing productivity. A study on the effects of habitat fragmentation on *Dianthus deltoides* L. in small forested areas has been shown to result in lowered seed set in plants due to changes in pollinators abundance and composition (Jennersten 1988). Furthermore, small and fragmented patches lack habitat heterogeneity to support abundant pollinators (Bawa 1990). This clearly shows plant–pollinator interactions are highly threatened in a fragmented landscape than a large and connected landscape. Consequently, this urges the need for the conservation of several small conservation reserves to enhance ecological networks and support abundant pollinators. The consequences of plant–pollinator disruption are beyond plant productivity; for instance, it could alter gene flow and lead to inbreeding among isolated habitats (Bawa 1990).

Apart from the abundance and diversity of pollinators, however, the plant–pollinator functional diversity increases plant community persistence (Fontaine et al. 2006). This study elucidated that plant communities pollinated by the most functionally diverse pollinator assemblage contained about 50% more plant species than did plant communities pollinated by less-diverse pollinator assemblages. Moreover, positive plant–pollinator interactions were observed.

Generally, in most human-dominated landscapes, there is in-proportionally small remnant natural vegetation. Much of the environment is degraded and sacred with scanty vegetation found either in areas that are not suitable for agriculture or sacred grooves. Hence, ecological restoration practices are getting due attention to rehabilitate a degraded environment. However, pollination received little attention in due process. Thus, the emphasis given to restoring vegetation should also be given to pollinators, because plant–pollination networks are indispensable. A recent report by Kaiser-Bunbury et al. (2017) showed ecosystem restoration strengthens pollination network resilience and function. Ecological restoration and conservation project planning must consider plant-animal interactions.

Small remnant patches are vital for pollinator visitors vis-à-vis biodiversity sustainability. These pollinators enable the reproduction of plant species. A study revealed that habitat fragmentation significantly reduced the abundance of pollinators, production and quality of fruit, and the survival rate of seedlings of recruitment in small stands (Hermansen et al. 2017). Consequently, the mutualistic interactions, pollen-pollinator interactions, are strengthened and productivity is improved by considering small conservation reserves. A recent study revealed that small forest patches increase bee visits to flowers in an oil palm plantation landscape in Borneo (Power et al. 2021). Therefore, protecting remnant patches as conservation reserves would attract several pollinators, which in turn promotes pollination. In the long term, it would maintain and increase plant diversity of an ecosystem.

6.2 Dispersal

Gene flow within and among plant communities is largely shaped by seed dispersal. This further enhances connectivity between plant populations. Thus, seed dispersal is vital for the persistence and spread of plant population (Rogers et al. 2021). In addition to the diversity and richness of dispersal agents, numerous factors indirectly influence seed dispersal. For instance, land-use changes (e.g., habitat destruction/loss, habitat degradation, and habitat fragmentation), invasive species, over-exploitation, climate change, and species extinction could influence seed dispersal.

Habitat fragmentation could lead to the loss of biotic dispersers such as mammals and birds, which in turn could cause the loss of plant species and functional diversity by reducing seed dispersal and recruitment (Stevenson and Aldana 2008; Osuri et al. 2020; Rogers et al. 2021). Biotic dispersers are common in closed forests or plant communities for instance in the Ethiopian highlands, unlike their counterpart abiotic dispersers which are common in deserts and semi-deserts for instance in the Afar rangelands, Ethiopia. Although it might vary at spatial scales, dispersers' loss could have direct and indirect influences on plant population and communities. A recent experimental study on abiotic dispersal also acknowledged that habitat fragmentation affects local seed dispersal primarily through the creation of edges and changing dispersal distance and direction (Warneke et al. 2021).

The loss of dispersers could directly reduce the plant species richness and composition and cause a decline in sapling and seedling richness in a plant community, either by limiting the colonization of species from the regional species pool into a local community (Levin et al. 2003; Stevenson 2011) or by limiting immigration from nearby growing populations that could rescue a shrinking population (Shmida and Ellner 1984). Some biotic dispersers have a narrow niche in a landscape. As a result, if the landscape is extremely fragmented, the access to these dispersers is severed. This in turn interrupts the gene flow to the respective fragment.

The indirect impact of dispersers' loss could be manifested in dispersal trait shifts. For instance, the loss of large mammal dispersers might cause a shift to small animal dispersers, which in turn contract the seed dispersal distance (Markl et al. 2012). Trait shift, for instance, small-seeded tree species were dominant after the loss of large seed dispersers in North America (Cardoso da Silva and Tabarelli 2000). Furthermore, ecosystem functions could be disrupted because of dispersers' loss. The ultimate indirect effect of dispersers' loss could be the metapopulation collapse which is eventually followed by regional species decline and extinction (Halley et al. 2016).

Similar to the plant–pollinator interactions, several small fragments have a potential to hold a distinct set of plant–seed disperser interactions, thereby maintaining functional diversity in fragmented tropical landscapes. This was strengthened by a rare study in the tropical forests of India (Jain et al. 2021). Hence, establishing small conservation reserves to improve the success of seed dispersal through a fragmented landscape could increase ecological connectivity (Chap. 9).

Small reserves, trees scattered in farmlands, Church forests, and home gardens could be considered as seedscape (Beckman and Rogers 2013)—biotic and abiotic environment surrounding a seed that influences later stages of plant recruitment—for plant dispersal in a human-dominated landscape. However, vacant patches could also be colonized via seed dispersal, which implies uncolonized patches could also serve as habitat connectivity. Albeit the impact of landscape structure and habitat area might differ, small patches combined with scattered trees and scattered trees only play a crucial role for the persistence and spread of plant population (Cadavid-florez et al. 2020; da Rocha et al. 2021).

In general, habitat fragmentation causes seed dispersers' loss, which in turn decreases the probability of seeds reaching the suitable habitat, which is vital for recruitment. Human-dominated landscapes are characterized by high habitat fragmentation and transformation. Hence, the abundance of seed dispersers in such a landscape is minimal. In order to increase the probability of seeds reaching suitable habitats and increase species richness and diversity, the protection of smaller patches as small reserves could serve as seedscape and habitat connectivity. Nevertheless, great care must be applied to avoid the patches not being a dead-end for the dispersers and limit seed dispersal.

Overall, pollination and seed dispersal are an important component of an ecosystem. They play a significant role in keeping gene flow, maintaining ecosystem stability, and enhancing species diversity. Fragmented landscapes could support some very important pollinators and dispersers. Thus, these fragments should be considered in ecological/ecosystem restoration to improve plant-animal interactions in an ecosystem.

References

Bawa KS (1990) Plant-pollinator interactions in tropical rain forests. Annu Rev Ecol Syst 21:399–422

Beckman NG, Rogers HS (2013) Consequences of seed dispersal for plant recruitment in tropical forests : interactions within the seedscape. Biotropica 0:1–16. https://doi.org/10.1111/btp.12071

Cadavid-florez L, Laborde J, James D et al (2020) Landscape and Urban planning Isolated trees and small woody patches greatly contribute to connectivity in highly fragmented tropical landscapes. Landsc Urban Plan 196:103745. https://doi.org/10.1016/j.landurbplan.2020.103745

CaraDonna PJ, Burkle LA, Schwarz B et al (2020) Seeing through the static: the temporal dimension of plant—animal mutualistic interactions. Ecol Lett. https://doi.org/10.1111/ele.13623

Cardoso da Silva J, Tabarelli M (2000) Tree species impoverishment and the future flora of the Atlantic forest of northeast Brazil . Nature 404:72–74

Chen I-C, Hill JK, Ohlemüller R et al (2011) Rapid range shifts of species associated with high levels of climate warming. Science (80-) 333:1024–1026. https://doi.org/10.1126/science.1206432

Clements F (1916) Plant succession: an analysis of the development of vegetation. Carnegie Institution of Washington

da Rocha EG da, Brigatti E, Branda B et al (2021) Dispersal movement through fragmented landscapes : the role of stepping stones and perceptual range. Landsc Ecol 36:3249–3267. https://doi.org/10.1007/s10980-021-01310-x

Fontaine C, Dajoz I, Meriguet J, Loreau M (2006) Functional diversity of plant—Pollinator interaction webs enhances the persistence of plant communities. PLoS Biol 4:e1. https://doi.org/10.1371/journal.pbio.0040001

Gleason H (1926) The individualistic concept of the plant association. Society 53:7–26. https://doi.org/10.2307/2479933

Halley JM et al (2016) Dynamics of extinction debt across five taxonomic groups. Nat Commun 7:12283

Hermansen TD, Minchinton TE, Ayre DJ (2017) Habitat fragmentation leads to reduced pollinator visitation, fruit production and recruitment in urban mangrove forests. Oecologia 185:221–231. https://doi.org/10.1007/s00442-017-3941-1

Hillebrand H (2004) On the generality of the latitudinal diversity gradient. Am Nat 163:192–221

Jain A, Page NV, Rawat GS, Naniwadekar R (2021) Are fragments fruitful? A comparison of plant–seed disperser communities between fragments and contiguous forest in north-east India. Biotropica 53:1368–1378. https://doi.org/10.1111/btp.12985

Jennersten O (1988) Pollination in Dianthus deltoides (Caryophyllaceae): effects of habitat fragmentation on visitation and seed set. Conserv Biol 2:359–366

Kaiser-Bunbury CN, Mougal J, Whittington AE et al (2017) Ecosystem restoration strengthens pollination network resilience and function. Nature 000:1–5

Kerr JT, Pindar A, Galpern P et al (2015) Climate change impacts on bumblebees converge across continents. Science (80-) 349:177–180

Knight TM, Steets JA, Vamosi JC et al (2005) Pollen limitation of plant reproduction: ecological and evolutionary causes and consequences. Annu Rev Ecol Evol Syst 36:467–497

Lenoir J, Gégout J-C, Marquet PA et al (2008) A significant upward shift in plant species optimum elevation during the 20th century. Science (80-)

Levin S, Muller-Landau H, Nathan R, Chave J (2003) The ecology and evolution of seed dispersal: a theoretical perspective. Annu Rev Ecol Evol Syst 34:575–604

Librán-Embid F, Grass I, Emer C et al (2021) A plant–pollinator metanetwork along a habitat fragmentation gradient. Ecol Lett 24:2700–2712. https://doi.org/10.1111/ele.13892

Lozier JD (2017) A helping habitat for bumblebees. Nature 1–2. https://doi.org/10.1038/nature21897

Markl J, Schleuning M, Forget P et al (2012) Meta-analysis of the effects of human disturbance on seed dispersal by animals. Conserv Biol 26:1072–1081

Nakazawa T, Doi H (2012) A perspective on match/mismatch of phenology in community contexts. Oikos 121:489–495

Ollerton J (2017) Pollinator diversity: distribution, ecological function, and conservation. Annu Rev Ecol Evol Syst 48:353–376. https://doi.org/10.1146/annurev-ecolsys-110316-022919

Ollerton J, Winfree R, Tarrant S (2011) How many flowering plants are pollinated by animals? Oikos 120:321–326. https://doi.org/10.1111/j.1600-0706.2010.18644.x

Osuri A, Mendiratta U, Naniwadekar R et al (2020) Hunting and forest modification have distinct defaunation impacts on tropical mammals and birds. Front for Glob Chang 2:87

Pellissier L, Alvarez N, Guisan A (2012) Pollinators as drivers of plant distribution and assemblage into communities. In: Patiny S (ed) Evolution of plant–pollinator relationships. Cambridge University Press, Cambridge, UK

Pellissier L, Bråthen KA et al (2010) Species distribution models reveal apparent competitive and facilitative effects of a dominant species on the distribution of tundra plants. Ecography (Cop) 33:1004–1014

Power CC, Nielsen A, Sheil D (2021) Even small forest patches increase bee visits to flowers in an oil palm plantation landscape. Biotropica 00:1–13. https://doi.org/10.1111/btp.13023

Rasmont P, Franzen M, Lecocq T et al (2015) Climatic risk and distribution atlas of European bumblebees. BioRisk 10:1–246

Rogers HS, Donoso I, Traveset A, Fricke EC (2021) Cascading impacts of seed disperser loss on plant communities and ecosystems. Annu Rev of Ecol Evol Syst 52:641–666. https://doi.org/10.1146/annurev-ecolsys-012221-111742

Settele J, Bishop J, Potts SG (2016) Climate change impacts on pollination. Nat Plants 2:16092

Shmida A, Ellner S (1984) Coexistence of plant species with similar niches. Vegetatio 58:29–55

Stevenson P (2011) The abundance of large ateline monkeys is positively associated with the diversity of plants regenerating in Neotropical forests. Biotropica 43:512–519

Stevenson P, Aldana A (2008) Potential effects of ateline extinction and forest fragmentation on plant diversity and composition in the western Orinoco Basin, Colombia. Int J Primatol 29:365–377

Warneke CR, Caughlin TT, Damschen EI et al (2021) Habitat fragmentation alters the distance of abiotic seed dispersal through edge effects and direction of dispersal. Ecology 103:1–8. https://doi.org/10.1002/ecy.3586

Chapter 7
Ecological Restoration

"Everything is in and from the forest" syndrome.

Abstract Ecological restoration is the process of assisting the recovery of an ecosystem that has been degraded, damaged, or destroyed. Due to the various motivations for ecological restoration, the definitions might also vary. Understanding the historical composition and structure of an ecosystem helps achieve the goals of ecological restoration. Restoration projects through area exclosure started in Tigray, northern Ethiopia, in the 1980s. Most of the researches conducted in Ethiopia to monitor the progress of the area exclosures were focused on either species composition, diversity, and structure, or soil physico-chemical properties, particularly, soil organic carbon stock (SOCS). Nevertheless, since these variables might not provide holistic insights about the complexity and emergent properties of the target habitats, restoration ecology should consider a new paradigm to restore complexity at multiple scales, in particular of individual ecosystems and across landscapes. The important ecosystem component, plant-animal interactions, should receive the desired attention in future ecological restoration projects/studies.

Keywords Area exclosure · Area enclosure · Diversity · Species composition · SOC · Soil seed bank · Re-vegetation

7.1 The Quest for Restoration

As discussed in the previous chapters, human beings have extremely transformed the planet earth. The ever-increasing human population exacerbates the situation. Most of the natural vegetation is, thus, converted into other land-use types. Degradation of natural vegetation is more severe in developing countries in which their Gross Domestic Product (GDP) is dependent on natural resources. According to FAO (2001), degradation is pronounced in the African continent which often leads to deforestation. The former USA senator, Albert Gore (AKA Al Gore), perfectly

K. G. Yaynemsa, *Plant Biodiversity Conservation in Ethiopia*,
https://doi.org/10.1007/978-3-031-20225-4_7

stated the situation in Ethiopia as the forest cover of Ethiopia has decreased from 40 to 1% since the 1950s. Concurrently, the amount of rainfall has declined to the point where the country is rapidly becoming a wasteland (Gore 1992). Even though several conservation activities are happening at the national level, its contribution to restoring degraded environments is not as it is supposed to be. Thus, the quest for ecological restoration is still seeking urgent attention. The issue of climate change is also forcing scientists and policy-makers to look into the ecological restoration perspective.

7.2 Definition of Ecological Restoration

Ecological restoration is defined differently by scholars though most of them are closely related. The various definitions of ecological restoration often emanate from the various motivations for ecological restoration. These motivations include conserving biodiversity, enhancing ecosystem processes, counteracting climate change, providing ecosystem services, providing employment and job training, reconnecting with nature and experiential education, conserving cultural values, atoning for past damages, and complying with legislation (Holl 2020). Usually, ecological restoration is interchangeably used with rehabilitation.

The widely used and commendable definition of ecological restoration is "the process of assisting the recovery of an ecosystem that has been degraded, damaged, or destroyed" Society for Ecological Restoration Science and Policy Working Group (SER 2004). The goal is to return a degraded ecosystem to its historic trajectory by focusing on ecosystem health, integrity, and sustainability. On the other hand, rehabilitation is defined as "Actions that aim to reinstate a level of ecosystem functionality where ecological restoration is not sought, but rather the focus is on the provision of ecosystem goods and services" (SER 2004). Both in ecological restoration and rehabilitation, there are historical disturbances and these need to be recovered, restored, rehabilitated, managed, or re-vegetated (Fig. 7.1). Ecological restoration considers both spatial and temporal dynamics.

Understanding the historical composition and structure of an ecosystem is not a simple task. Unless and otherwise these are known, achieving the ecological restoration goal could be challenging. Furthermore, restoring to its historic trajectory seems far from attainable for several reasons. For example, if there is no record about the original composition and structure of the ecosystem to be restored, monitoring the success of ecological restoration would not be justified in the absence of baseline data. Even some species might be already extinct. Hence, these species have almost zero probability to be restored. However, biodiversity benefits the most from the ecological restoration. If biodiversity conservation is improved, ecosystem services would be improved too to benefit the socio-economic status of the people.

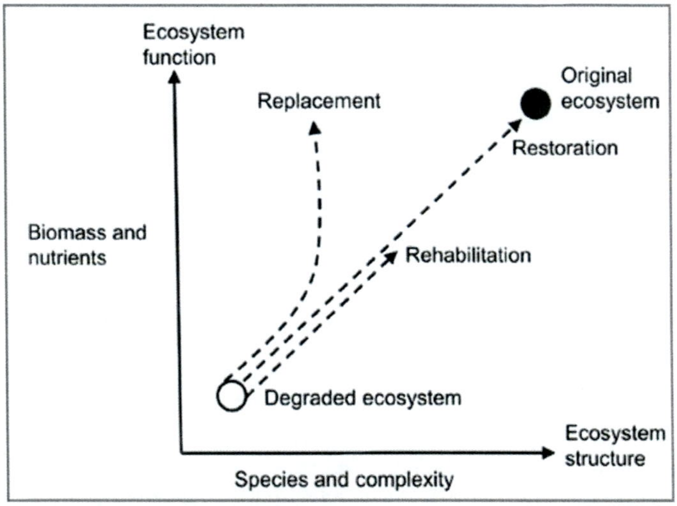

Fig. 7.1 Classic view of restoration (Holl 2020)

7.3 Novel Ecosystem Restoration

Human-induced factors exerted significant pressure on the earth's ecosystem. This has transformed a significant portion of the terrestrial ecosystem. As a result, biodiversity, ecosystem functions, and productivity are severely affected. Furthermore, the introduction of non-native species to an ecosystem is becoming common in different biomes of the world. Consequently, several ecosystem types have emerged over the due process. Novel ecosystems are the results of such ecosystem processes.

The epicenter of the novel ecosystems concept is the combination of new species composition and the magnitude of human interference to maintain the newly emerging ecosystem (Hobbs et al. 2006). The new species composition can alter ecosystem functions, and the human factor could not be sustainable for the maintenance of the novel ecosystem. The details about novel ecosystems are discussed in Chap. 8.

7.4 Ecological Restoration Projects in Ethiopia

The northernmost of Ethiopia is environmentally degraded due to historical reasons. Several devastating wars, including the recent Ethio-Eritrea war, have been undertaken in this region. It is obvious that such recurrent phenomena have a devastating impact on the environment. Land transformations, deforestation, and drought have been also aggravating environmental degradation across Tigray in particular and Ethiopia at large. Consequently, the region is covered with scanty remnant natural vegetation. Restoration projects through area exclosure started in Tigray, northern Ethiopia, in the 1980s (Nedessa et al. 2005). To restore the degraded ecosystem, a few approaches have been applied. Area exclosure sometimes interchangeably used with area enclosure though have different meanings is one of the approaches which is getting recognition in recent years at the national level. Area exclosure is discussed in the following section briefly.

7.4.1 Area Exclosure

Some authors are confused to differentiate area exclosure and area closure. A reader might encounter either of these terminologies for the same approach on the ground. To clarify this confusion, Aerts et al. (2009) published a note on the definitions of these terminologies. Accordingly, area enclosures are "areas surrounded by walls, objects, or other structures" often designed to protect physical disturbances such as livestock trampling/grazing or unwanted pollinators in experimental trials. On the other hand, area exclosures are "areas from which unwanted animals, etc., are excluded." The purpose of both exclosures and enclosures is, however, more or less similar. For any area or activity that involves excluding unwanted species or practices from environmentally degraded sites, the authors recommended the term area exclosure. Hence, area exclosure will be used throughout this book.

Area exclosure is a mechanism of assisting the recovery of a degraded ecosystem. The primary goal is to restore biodiversity and ecosystem services. These further benefit the community through both socio-economic and cultural services. As mentioned earlier, the experience of area exclosure is about three decades, which is relatively young. It is premature to conclude that a degraded environment is fully restored to its original state within this short period. Nevertheless, sharing the progress of the area exclosures could help researchers, decision-makers, and policy-makers to envisage the pros and cons.

Most of the researches conducted in Ethiopia to monitor the progress of the area exclosures were focused on either species composition, diversity, and structure, or soil physico-chemical properties, particularly, soil organic carbon stock (SOCS). Nevertheless, since these variables might not provide holistic insights about the complexity and emergent properties of the target habitats, restoration ecology should consider a new paradigm to restore complexity at multiple scales, in particular of individual ecosystems and across landscapes (Bullock et al. 2021). The multi-scale variables might include β diversity, heterogeneity among habitat patches, and connectivity. The important ecosystem component, plant-animal interactions, should receive the desired attention in future ecological restoration projects/studies. These include seed dispersal, pollination, and herbivory. I will briefly explain the importance of area exclosures in Ethiopia from a biodiversity perspective (floristic composition, structure, diversity, etc.), soil seed bank, and soil physico-chemical properties (e.g., soil organic matter, macro- and micronutrients, bulk density, etc.).

7.4.1.1 Species Composition, Structure, and Diversity

After the implementation of area exclosure projects in Tigray, some comparative studies were undertaken to understand the progress of the restoration projects. A study in the Tselemti district of western Tigray, Ethiopia, elucidated significantly higher species richness and density in 10-year-old area exclosure than five-year-old area (+10 species) exclosures and the open grazed areas (+20 species) (Gebregerges et al. 2017). This clearly implies age is a determinant factor in area exclosure besides other factors. Even though the woody species composition was higher in the open grazing land than the area exclosure, herbaceous biomass and diversity were by far higher in the area exclosure than the open grazing land (Fig. 7.2a) (Haftay et al. 2013). Similarly, regardless of area exclosure age, Mekuria and Veldkamp (2012) showed higher species richness, diversity, and biomass in area exclosure than open communal land (Fig. 7.2b). Others also confirmed high species richness and diversity in area exclosures than adjacent communal grazing areas in different parts of Ethiopia (Mengistu et al. 2005; Mekuria et al. 2018; Sinore and Doboch 2021). In most area exclosures, however, almost no tree species were recorded. For example, Tewolde-Berhan et al. (2003) recorded only two tree species with a single individual each on a 25-year-old area closure. For obvious reasons, a botanist/ecologist cannot expect a forest over a period of three decades from a bare land. Nevertheless, this is really below the expectations. On the contrary, Mekuria et al. (2018) recorded relatively high tree species richness in other area closure.

7.4.1.2 Soil Organic Matter

Soil is the major reservoir of organic carbon (Gebrehiwot et al. 2018). Edaphic variables, physico-chemical, are excellent indicators of ecosystem quality. An ecosystem with good natural vegetation cover is endowed with good soil quality such as higher soil organic matter (SOM) than the degraded environment and vice versa. Soil organic matter is either fresh or partially decomposed matter that comprises plant and animal detritus, soil microorganisms, and humus. Several comparative studies of area exclosures and their adjacent communal grazing lands have proved this issue.

High above-ground carbon stocks in the 15-year-old area exclosures (9.08 ± 1.44 Mgha^{-1}) attributed to intact woody and grass vegetation as well as substantial litter cover (>20% of the total biomass) (Araya et al. 2020) (Fig. 7.3a). A comparative study by Corral-NuNez et al. (2014) revealed the SOM content of farmland soils is small, ranging from 2.1 to 2.9%, while area exclosures showed significant recovery after 20 years of protection, with SOM contents ranging from 2.6 to 5.6%. Furthermore, higher soil organic carbon stock was reported in naturally regenerated area exclosures than assisted area closures in northern Ethiopia (Hishe et al. 2020). Despite the increasing trend of soil organic carbon with area exclosure age, Aynekulu et al. (2017) reported a nonsignificant difference in soil organic carbon stock between the area exclosures and adjacent open grazed rangelands in Borana lowlands, southern Ethiopia. Mekuria et al. (2018) argued that slope position has

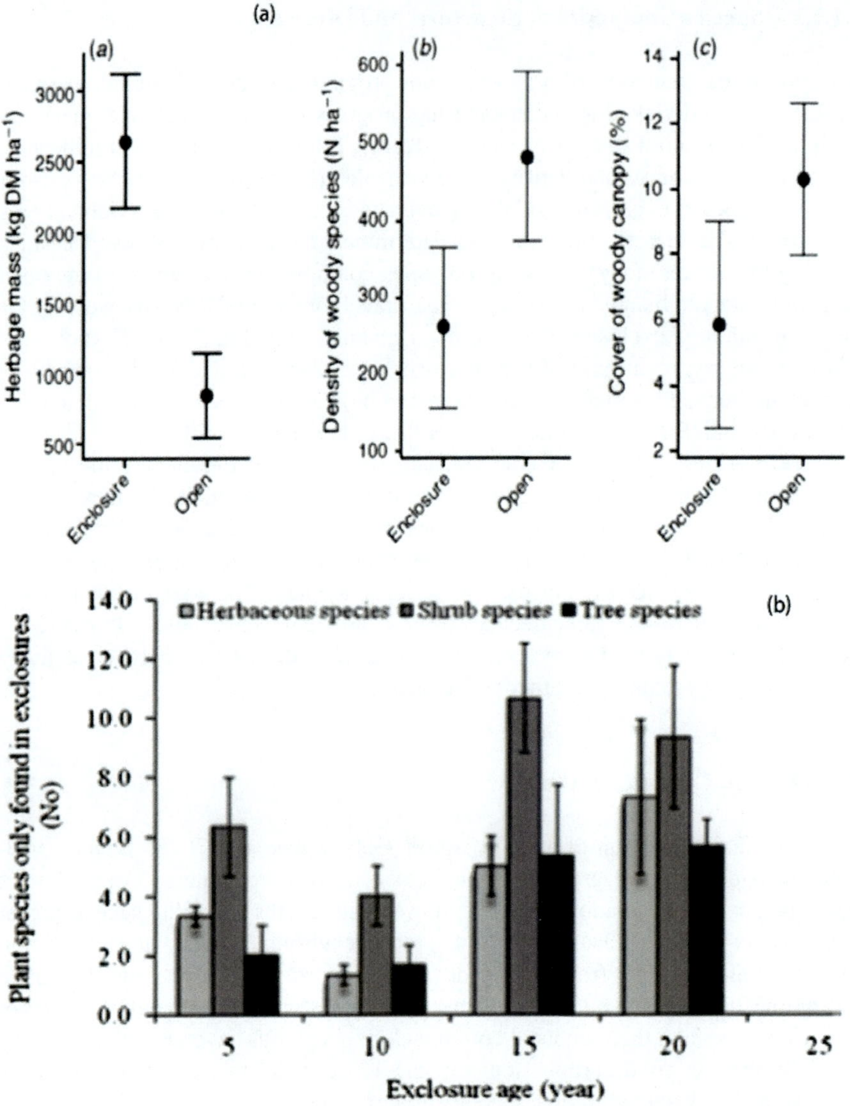

Fig. 7.2 Comparison of area exclosure and other land uses. **a** Herbaceous mass, the density of woody species, and cover of the woody canopy (%) of area exclosures and open-access grazing areas (Haftay et al. 2013) reproduced with permission from CSIRO Publishing and **b** average number of herbaceous, shrub, and tree species observed only in exclosures (Mekuria and Veldkamp 2012) reproduced with permission from John Wiley and Sons

a significant impact on soil organic carbon stock and other soil properties such as total Nitrogen. Although the reports claimed a promising outcome of an area exclosures, it is far from a successfully restored environment. Hence, policy-makers and decision-makers should be aware that there are several factors to be considered before applying area exclosure. These include management system (natural or assisted) area exclosures and the slope position of the site to be selected for area exclosures.

7.4.1.3 Soil Seed Bank

Area exclosures are also the hub for soil seeds. Soil seed bank is the crux of natural regeneration in an ecosystem. The soil seed bank and dispersed propagules helped in relatively quick recovery of the degraded environment. Higher soil seed bank density in area exclosure than the neighboring open grazed land was reported in the central and northern parts of Ethiopia (Mengistu et al. 2005). Mohammed and Denboba (2020) also compared the density, richness, and density of soil seed banks in area exclosures and open grazing land. They found significantly higher species richness, diversity, and density in the area exclosure of different years old than open grazing lands. This implies that area exclosures could play a vital role in restoring a degraded environment via their soil seed bank. However, there is a worrisome issue in some area exclosures with regard to the soil seed bank composition. Mengistu et al. (2005) showed *Acacia saligna* (Labill.) H.L.Wendl., which is a presumably invasive plant, is one of the dominant species in the soil seed bank. Thus, curious monitoring and selection of plant species to assist restoration of a degraded environment should be inevitable. If not, such species could lead to unprecedented phenomena such as ecosystem modification within a short period.

7.4.2 Plant Species Selection for Re-vegetation

Apart from area exclosures, Ethiopia has been working on the re-vegetation of degraded environment aiming at improving vegetation, particularly forest, cover. This is associated with the Climate Resilient Green Economy (CRGE) policy that was initiated by the Federal Democratic Republic of Ethiopia (FDRE) in 2011. It was under the leadership of the Prime Minister's office, Environmental Protection Authority, and the Ethiopian Development Research Institute. Mass plantation of saplings has been started especially in the Ethiopian Millennium (2008) even by establishing Millennium parks almost in every region of the country. Since then, plantations have been continued on yearly basis. In 2019, the country boasted to have planted 4 billion saplings just in a single day although the confirmation and monitoring approach are hardly stated.

There are three major flaws related to the mass plantation on yearly basis. First, there is no standard system to monitor the progress of the once planted saplings. Hence, it is challenging to track their growth stage, how many of them are surviving,

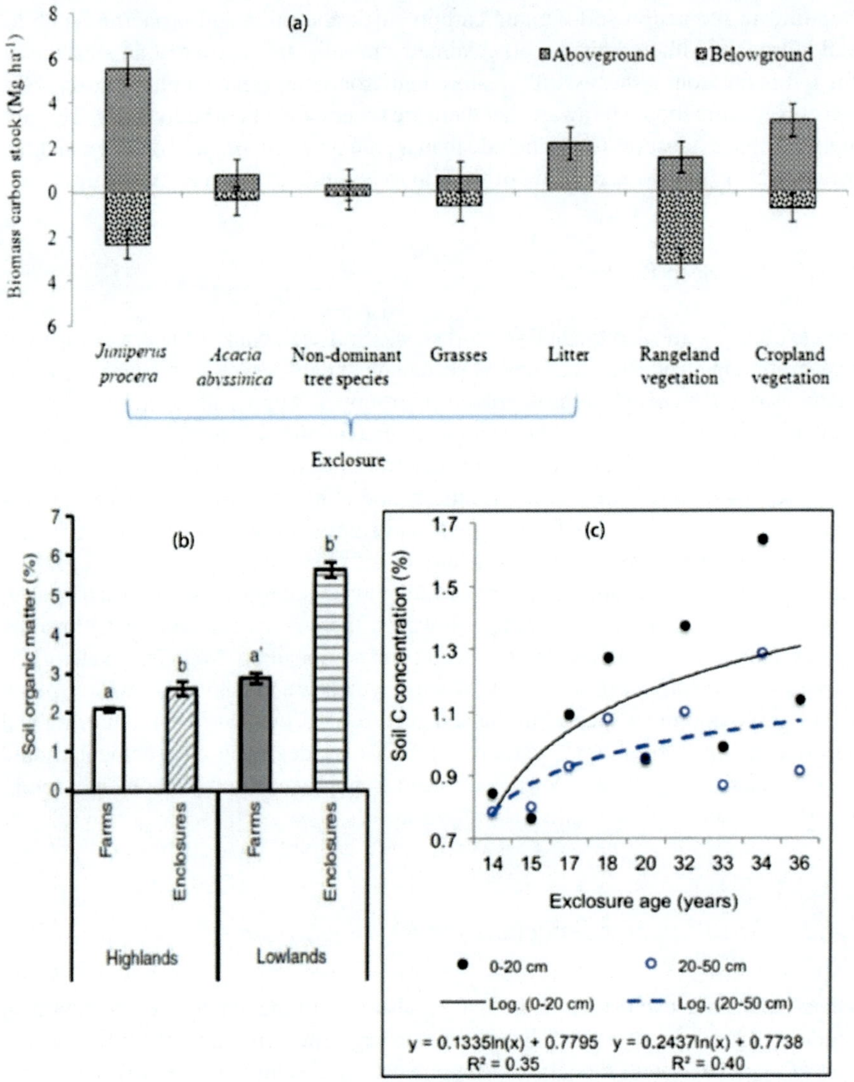

Fig. 7.3 Comparison of area exclosure and other land uses. **a** Above- and below-ground biomass carbon stock contributed by different plant species in major land-use systems of the Gergera watershed (Araya et al. 2020) reproduced with permission from Elsevier, **b** current soil organic matter content in farms and exclosures (Corral-NuNez et al. 2014) reproduced with permission from John Wiley and Sons, and **c** soil carbon content at 0–20 and 20–50 cm depths at different exclosure ages (Aynekulu et al. 2017) reproduced with permission from Elsevier

and what are the main threats to these plants. Even though I could not find a published report on this issue, presumably most of the planted saplings do not even survive until the next rainy season, especially in regions that receive low annual rainfall. The second drawback is the composition of plants selected for re-vegetation.

Most of the plants selected for re-vegetation are exotic tree species. Some of the exotic plants selected for re-vegetation even turned into invasive. For example, *A. saligna* and *Acacia mearnsii* De Wild. are planted in degraded environments and later turned either into invasive or potentially invasive plants. Both of them are invasive in other countries such as South Africa. Some of the exotic species commonly encountered in plantations are shown in Table 7.1 and Fig. 7.4. It does not mean that exotic species are irrelevant in restoring a degraded environment; however, prior assessment of their impact on the environment must be conducted. For example, their functional ecology such as invasive behavior and impact on soil physico-chemical properties should be assessed before mass planting action is taken. Few studies have made a comparative study between exotic and indigenous plants.

A comparison of understory vegetation and soil fertility in plantations and adjacent natural forests in the Ethiopian highlands revealed that indigenous vegetation has higher soil total Nitrogen, available Phosphorus, and exchangeable Calcium than the exotic plantations (Michelsen et al. 1996). In terms of the herbaceous species, the *Eucalyptus* and *Pinus* plantations hosted almost similar richness and diversity to the natural forest though low richness and diversity were recorded in the nine-year-old *Cupressus* plantations. Similarly, a study in the central highlands of Ethiopia showed indigenous forests entertained higher soil organic carbon, total Nitrogen, and high tree diversity (α-diversity, β-diversity, and γ-diversity) than the exotic plantations (Tekalign et al. 2017). An experimental study in northwestern Ethiopia also proved

Table 7.1 List of some exotic plant species planted to restore a degraded environment in Ethiopia

No	Species name	Family	Growth habit
1	*Grevillea robusta* A.Cunn. ex R.Br	Proteaceae	Tree
2	*Acacia melanoxylon* R.Br	Fabaceae	Tree
3	*Acacia saligna* (Labill.) H.L.Wendl	Fabaceae	Shrub/tree
4	*Cupressus lusitanica* Mill	Cupressaceae	Tree
5	*Cajanus cajan* (L.) Huth	Fabaceae	Shrub
6	*Eucalyptus camaldulensis* Dehnh	Myrtaceae	Tree
7	*Eucalyptus globulus* Labill	Myrtaceae	Tree
8	*Leucaena leucocephala* (Lam.) de Wit	Fabaceae	Tree
9	*Casuarina equisetifolia* L.	Casuarinaceae	Tree
10	*Schinus molle* L.	Anacardiaceae	Tree
11	*Pinus sylvestris* L.	Pinaceae	Tree
12	*Jacaranda mimosifolia* D.Don	Bignoniaceae	Tree
13	*Sesbania sesban* (L.) Merr	Fabaceae	Shrub/tree
14	*Acacia decurrens* (J.C.Wendl.) Willd	Fabaceae	Tree

Fig. 7.4 Exotic species observed in and around re-vegetated areas. Seedlings of **a** *A. saligna*, **b** *A. mearnsii,* and **c** *S. sesban* to be deployed to the field

the importance of indigenous plants over exotic plants in restoration practices (Abiyu et al. 2017). These authors reported that the indigenous plant, *Olea europaea* subsp. *cuspidata* (Wall. & G.Don) Cif., regenerated only in the previously existing native non-browsing species. Furthermore, the soil nutrients are richer in these pre-existing native non-browsing species than the sown and planted exotic plants. This is an alert that priorities should be given to either indigenous or exotic species after a long-term trial of their impact on the indigenous plants and soil physico-chemical properties as well.

The third one is the "*Everything is in and from the forest*" syndrome. It is a perception that, if forests are restored, everything would be restored, and the ecosystem values and services obtained from it would be improved. Often, tree plantation campaigns in Ethiopia promote planting saplings on open biomes where forests might not grow historically. Most of the time policy-makers in Ethiopia did not consult experts (personal observation), and they perceive open biomes are degraded ecosystems. This primarily neglects or replaces the ecological and economic importance of open ecosystems (Bond 2019; Silveira et al. 2021) such as grasslands, shrublands, and woodlands. This claim is strengthened by the national reports, which boasts about forest cover improvement but say nothing about open ecosystems. These practices destroy the habitats of plants and animals adapted to open ecosystems and threaten the livelihoods of people dependent on those ecosystems to produce wild game and domestic livestock (Fleischman et al. 2020). To avoid this drawback conservation, experts might consult potential natural vegetation maps (Chap. 3).

Studies on dynamics of land cover change indicated grasslands are being converted into agricultural lands in Ethiopia (Worku et al. 2016; Gebremedhin et al. 2018; Gebrehiwot et al. 2021). In addition, a Human Appropriation Net Primary Production (HANPP) analysis claimed that Ethiopia's grazing intensity is among the highest observed globally, with HANNP of about 68% (Grabher 2021). This implies grasslands are threatened. Neglecting the open biomes is also reflected globally, even by the UN Decade on Ecosystem Restoration (Silveira et al. 2021). Different lands or ecosystems deliver different purpose; i.e., forests could not deliver everything. Hence, restoration projects in Ethiopia should also consider open ecosystems.

Overall, restoration projects through enrichment often must at least pass through certain phases. These include restoration site selection, understanding the potential natural vegetation [reference points], plant species selection and material [seed] preparation, and management and regular surveys. Furthermore, a national ecological restoration guideline might resolve the ambiguity of restoration-related issues.

References

Abiyu A, Teketay D, Glatzel G et al (2017) Restoration of degraded ecosystems in the Afromontane highlands of Ethiopia : comparison of plantations and natural regeneration. South J For Sci ISSN 2620. https://doi.org/10.2989/20702620.2016.1254917

Aerts R, Nyssen J, Haile M (2009) On the difference between "exclosures" and "enclosures" in ecology and the environment. J Arid Environ 73:762–763. https://doi.org/10.1016/j.jaridenv.2009.01.006

Araya T, Khamzina A, Gebresamuel G, Amelung W (2020) Terrestrial carbon stocks following 15 years of integrated watershed management intervention in semi-arid Ethiopia. CATENA 190:104543. https://doi.org/10.1016/j.catena.2020.104543

Aynekulu E, Mekuria W, Tsegaye D et al (2017) Long-term livestock exclosure did not affect soil carbon in southern Ethiopian rangelands. Geoderma 307:1–7. https://doi.org/10.1016/j.geoderma.2017.07.030

Bond WJ (2019) Open ecosystems: ecology and evolution beyond the forest edge. Oxford University Press, UK

Bullock JM, Fuentes-montemayor E, Mccarthy B, et al (2021) Future restoration should enhance ecological complexity and emergent properties at multiple scales. Ecography (Cop) 44:1–11.https://doi.org/10.1111/ecog.05780

Corral-NuNez G, Opazo-Salazar D, Gebresamuel G et al (2014) Soil organic matter in Northern Ethiopia, current level and predicted trend : a study case of two villages in Tigray. Soil Use Manag 30:487–495.https://doi.org/10.1111/sum.12157

FAO (2001) Global forest resources assessment 2000. FAO, Rome, Italy

Fleischman F, Basant S, Chhatre A et al (2020) Pitfalls of tree planting show why we need people-centered natural climate solutions. Bioscience 70:947–950. https://doi.org/10.1093/biosci/biaa094

Gebregerges T, Tessema ZK, Birhane E (2017) Effect of exclosure ages on woody plant structure, diversity and regeneration potential in the western Tigray region of Ethiopia. J for Res. https://doi.org/10.1007/s11676-017-0512-6

Gebrehiwot K, Desalegn T, Woldu Z et al (2018) Soil organic carbon stock in Abune Yosef afroalpine and sub-afroalpine vegetation, northern Ethiopia. Ecol Process 7:6. https://doi.org/10.1186/s13717-018-0117-9

Gebrehiwot K, Teferi E, Woldu Z et al (2021) Dynamics and drivers of land cover change in the Afroalpine vegetation belt: Abune Yosef mountain range, Northern Ethiopia. Environ Dev Sustain 23:10679–10701. https://doi.org/10.1007/s10668-020-01079-0

Gebremedhin H, Gebresamual G, Abadi N et al (2018) Conversion of communal grazing land into arable land and its impacts on soil properties and vegetation cover. Arid L Res Manag 32:236–252. https://doi.org/10.1080/15324982.2017.1406412

Gore A (1992) Earth in the balance: forging a new common purpose. Earthscan, UK

Grabher H (2021) HANPP trajectories for Ethiopia reveal recent agricultural efficiency gains but high grazing intensity. Environ Dev Sustain 23:5277–5296. https://doi.org/10.1007/s10668-020-00814-x

Haftay H, Yayneshet T, Animut G, Treydte AC (2013) Rangeland vegetation responses to traditional enclosure management in eastern Ethiopia. Rangel J 35:29–36. https://doi.org/10.1071/RJ12054

Hishe H, Giday K, Soromessa T et al (2020) Should we leave nature unattended or assist through enrichment to foster climate change mitigation? Exclosure management in the highlands of Ethiopia. Environ Manage. https://doi.org/10.1007/s00267-020-01259-8

Hobbs RJ, Arico S, Aronson J et al (2006) Novel ecosystems : theoretical and management aspects of the new ecological world order. Glob Ecol Biogeogr 15:1–7. https://doi.org/10.1111/j.1466-822x.2006.00212.x

Holl KD (2020) Primer of ecological restoration. IslandPress, Washington, USA

Mekuria W, Veldkamp E (2012) Restoration of native vegetation following exclosure establishment on communal grazing lands in Tigray. Appl Veg Sci 15:71–83. https://doi.org/10.1111/j.1654-109X.2011.01145.x

Mekuria W, Wondie M, Amare T et al (2018) Restoration of degraded landscapes for ecosystem services in North-Western Ethiopia. Heliyon. https://doi.org/10.1016/j.heliyon.2018.e00764

Mengistu T, Teketay D, Hulten H, Yemshaw Y (2005) The role of enclosures in the recovery of woody vegetation in degraded dryland hillsides of central and northern Ethiopia. J Arid Environ 60:259–281. https://doi.org/10.1016/j.jaridenv.2004.03.014

Michelsen A, Lisanework N, Friis I, Holst N (1996) Comparisons of understorey vegetation and soil fertility in plantations and adjacent natural forests in the Ethiopian highlands. J Appl Ecol 33:627–642. https://doi.org/10.2307/2404991

Mohammed SA, Denboba MA (2020) Study of soil seed banks in ex-closures for restoration of degraded lands in the central rift valley of Ethiopia. Sci Rep 10:956. https://doi.org/10.1038/s41598-020-57651-1

Nedessa B, Ali J, Nyborg I (2005) Exploring ecological and socio-economic issues for the improvement of area enclosure management: a case study from Ethiopia

SER (2004) The SER primer on ecological restoration. Society for Ecology Restoration International, Washington, DC

Silveira FAO, Ordóñez-Parra CA, Moura LC et al (2021) Biome awareness disparity is BAD for tropical ecosystem conservation and restoration. J Appl Ecol 00:1–9. https://doi.org/10.1111/1365-2664.14060

Sinore T, Doboch D (2021) Area exclosure effects on soil properties and woody species diversity in Amelake sub-watershed, Southern Ethiopia. Int J River Basin Manag. https://doi.org/10.1080/15715124.2020.1843469

Tekalign M, Van MK, Aerts R et al (2017) Effects of biodiversity loss and restoration scenarios on tree-related ecosystem services. Int J Biodivers Sci Ecosyst Serv Manag 13:434–443. https://doi.org/10.1080/21513732.2017.1399929

Tewolde-Berhan S, Mitlöhner R, Muys B, Haile M (2003) Vegetation improvement in communal closed areas in Tigray, Ethiopia. In: XII World Foretry Congress. Quebec City, Canada, pp 1–8

Worku T, Tripathi SK, Khare D (2016) Analyses of land use and land cover change dynamics using GIS and remote sensing during 1984 and 2015 in the Beressa Watershed Northern Central Highland of Ethiopia. Model Earth Syst Environ 2:1–12. https://doi.org/10.1007/s40808-016-0233-4

Chapter 8
Novel Ecosystem

Abstract Scholars define novel ecosystems differently. A novel ecosystem is a unique assemblage of biota and environmental conditions that is the direct result of intentional or unintentional alteration by humans, i.e., human agency, sufficient to cross an ecological threshold that facilitates a new ecosystem trajectory and inhibits its return to a previous trajectory regardless of additional human intervention. There are several questions raised related to the emerging concept, i.e., novel ecosystem. These include the following: (1) What makes a novel ecosystem different from other ecosystems? (2) Does this concept contradict the very basic aim of ecological restoration? (3) Is novel ecosystem use and service-oriented? (4) Does this encourage the introduction of non-native species? (5) Are all sorts of man-induced factors considered as an agency in the novel ecosystem? (6) What is the reference point for threshold? (7) Is the focus unique species, richness, or new combination and abundance? and (8) Are novel ecosystems persistent and resilient? These are robustly discussed in this chapter.

Keywords Ecological threshold · Invasive species · Novel ecosystem · Self-sustaining · Species composition

8.1 Introduction

Novel ecosystem is a trending phrase since its first use in a publication in the last 1990s (Chapin and Starfield 1997). However, several suspicious critics emerged after the publication of Hobbs et al. (2006). Several ecologists and conservation experts frequently raise the following questions.

- What is a novel ecosystem?
- What makes a novel ecosystem different from other ecosystems?
- Does this concept contradict with the very basic aim of ecological restoration?
- Is novel ecosystem use and service-oriented?
- Does this encourage the introduction of non-native species? etc.

K. G. Yaynemsa, *Plant Biodiversity Conservation in Ethiopia*,
https://doi.org/10.1007/978-3-031-20225-4_8

Bringing these and other questions to academia is reasonable and must be settled in order to integrate the concept into ecosystem management and restoration practices. This begins with having an ambiguous definition of the term.

Scholars define novel ecosystems differently (Tables 8.1 and 8.2) although the concept is extremely related. The highly cited definition of the novel ecosystem is "A novel ecosystem is a system of abiotic, biotic and social components (and their interactions) that, under human influence, differ from those that prevailed historically, tending to self-organize and manifest novel qualities without intensive human management" (Hobbs et al. 2013). However, a year later after a relatively explained definition of novel ecosystems emerged. The new definition is "A novel ecosystem is a unique assemblage of biota and environmental conditions that is the direct result of intentional or unintentional alteration by humans, i.e., human agency, sufficient to cross an ecological threshold that facilitates a new ecosystem trajectory and inhibits its return to a previous trajectory regardless of additional human intervention." The resulting ecosystem must also be self-sustaining in terms of species composition, structure, biogeochemistry, and ecosystem services. "A defining characteristic of a novel ecosystem is a change in species composition relative to ecosystems present in the same biome prior to crossing a threshold" (Morse et al. 2014). In most of the definitions, the influence of human beings leading to novel ecosystem formation is signified. There are no arguments about the contribution of anthropogenic factors to the ecosystem. Rather, the arguments are about restoring or managing the ecosystems formed because of anthropogenic disturbances. Furthermore, its application is difficult because of the complexity to identify if an ecosystem is novel or not (Murcia et al. 2014).

The arguments about a novel ecosystem start from its definition. Most scholars claim that the definition of a novel ecosystem is a mere modification of the ecosystem definition. For instance, self-organization is a typical characteristic of an ecosystem. Furthermore, the technical terms (human agency, threshold, species composition, and self-sustaining) need additional explanation (Morse et al. 2014). Questions related to the definition include the following: (1) Are all sorts of man-induced factors considered as an agency in the novel ecosystem? (2) What is the reference point for threshold? (3) Is the focus unique species, richness, or new combination and abundance? and (4) Are novel ecosystems persistent and resilient? These and other questions should be clarified to help policy- and decision-makers. These are discussed in the following sub-sections.

8.1.1 Human Factor

Human beings have been altering the ecosystem for millions of years. Literally, every ecosystem on earth is affected by either direct or indirect anthropogenic disturbances. As stated in Chap. 1, researchers estimated up to 78% of the planet today is considered anthropogenic biomes (anthromes) (Ellis et al. 2010). For instance, the Ox-plow in Ethiopia, in its current form as a dominant tool, appears in rock painting

Table 8.1 Definitions of novel ecosystem

No	Definitions	Author
1	A novel ecosystem is a system of abiotic, biotic, and social components (and their interactions) that, by virtue of human influence, differ from those that prevailed historically, having a tendency to self-organize and manifest novel qualities without intensive human management	Hobbs et al. (2013)
2	Novel ecosystems are self-maintaining ecosystems that support species assemblages without historical precedent	Fehr et al. (2020)
3	Novel combinations of species that could alter ecosystem processes and a genesis in human action or species compositions and relative abundances that have not occurred previously within a given biome	Hobbs et al. (2006)
4	Novel ecosystems, as defined here, are those composed of reclaimed stands growing on anthropogenically reconstructed soil	Norris et al. (2013)
5	A novel ecosystem is a unique assemblage of biota and environmental conditions that is the direct result of intentional or unintentional alteration by humans, i.e., human agency, sufficient to cross an ecological threshold that facilitates a new ecosystem trajectory and inhibits its return to a previous trajectory regardless of additional human intervention. The resulting ecosystem must also *be self-sustaining in terms of species composition, structure, biogeochemistry, and ecosystem services. A defining characteristic of a novel ecosystem is a change in species composition relative to ecosystems present in the same biome prior to crossing a threshold*	Morse et al. (2014)
7	Regenerating forests on degraded lands, composed largely of non-native species and exhibiting multiple successional pathways	Lugo and Helmer (2004)
8	Novel forests dominated by introduced species maintain ecosystem services after decline of native tree species	Mascaro et al. (2012)
9	Invasion by non-native species (e.g., suspension-feeding bivalves, crustaceans, and fish), stocking and accidental release of hybrid and non-native species	Strayer (2010)

dating as far back as 500 AD (McCann 2017). Hence, if every human impact on the ecosystem is counted, we might designate every ecosystem on planet earth as a novel ecosystem. Even though the human factor is prominent in the assembly of an ecological community, what is its magnitude to call a driver of a novel ecosystem?

Morse et al. (2014) clearly indicated that it is not sound to consider direct or indirect factors (e.g., climate change) that affect all ecosystems change over time. Rather, it should be the human factors that distinguish novel ecosystems from other ecosystems. Furthermore, the ecosystem trajectories must be initiated by the human factor directly to designate it novel ecosystem (Lugo and Helmer 2004). Moreover,

Table 8.2 Some novel ecosystem types studied globally

No	Novel ecosystem type	Author
1	Mixed exotic-native tropical forests establishing on abandoned pasture and cropland in Puerto Rico	Lugo and Helmer (2004)
2	Nearshore ocean floors dominated by the exotic invasive alga *Caulerpa*	Davis et al. (1997)
3	A grassland in Colorado (USA) composed of a unique assemblage of grass species planted by managers to restore an abandoned gravel pit	Seastedt et al. (2008)
4	Urban regions with heavily altered abiotic conditions and large numbers of exotic species (Kowarik 2011)	Lindenmayer et al. (2008)
6	Sand and gravel mining create novel ecosystems	Orgensen et al. (2021)
7	Urban forest remnants of the Atlantic Forest biome, in the city of Rio de Janeiro	Solórzano et al. (2021)
8	Hawai'i mangroves as a case study	Lewis and Granek (2021)
9	Post-industrial areas: The case of a former brown coal mine in Søby, Denmark	Conradi et al. (2021)

it should be noted that regular human management is unnecessary to maintain novel ecosystems.

The deliberate introduction of *Prosopis juliflora* (Sw.) DC in the 1970s to restore the arid areas in eastern Ethiopia is a typical example of direct initiation. This species is now invasive in Ethiopia, which transformed the arid ecosystem. According to Shiferaw et al. (2019a, b), the estimated rate of invasion is 31,127 ha per year. Furthermore, mass plantation of exotic species that are mobilized in the notion of "Climate Resilient Green Economy" or "green legacy" might be leading to the creation of novel ecosystems of Ethiopia. The age-old plantations and dynamics of land cover change are some drivers of novel ecosystems around the world in general and Ethiopia in particular.

8.1.2 Self-sustaining (Persistence and Resilience)

Ecosystems entertain extremely complex interactions of biotic and abiotic components. The ability of an ecosystem to remain resilient and persistent is known as self-sustaining or self-organization. If self-sustaining/self-organization is achieved in the absence of significant human intervention, then the ecosystem is a novel ecosystem (Hobbs et al. 2013; Morse et al. 2014) and host species assemblages without historical precedents (Fehr et al. 2020). The term without intensive or significant human intervention is often ambiguous and unmeasurable. Nevertheless, Morse et al. (2014) implied that ecosystem stability is influenced by time scale and management practices. As a result, stakeholders might consider (Lugo and Helmer 2004) 10–100 years (Ferraz et al. 2020) time scale to detect if the novel ecosystem is self-sustaining or not. This is because a short time scale will cause confusion with seasonality, population/community dynamics, and random variation (Levin 1992), and a long time scale will cause confusion with natural succession and ecosystem turnover (Levin 1992).

The staunch opponent of a novel ecosystem concept (Simberloff et al. 2015) criticized that there is no proof that novel ecosystems are self-sustaining bitterly describing "the novel-ecosystem label can serve as a Get Out of Jail Free card for companies or individuals trying to avoid investing in research, mitigation or restoration by claiming they are producing novel ecosystems that will provide ecosystem services."

8.1.3 Species Composition

Based on their origin, species could be native or exotic. If we consider species dynamics, the species might be categorized as casual, naturalized, and invasive species. These species might also be casual species that could not be considered in novel ecosystems because they do not form a self-sustaining population (Fehr et al. 2020). Exotic species, whether they are invasive or non-invasive, often create a new composition of species with different abundance from the historical plant community. This forms a novel ecosystem (Hobbs et al. 2006; Morse et al. 2014). The species composition and abundance of a novel ecosystem are considered unique if it harbors species composition and abundance that barely exist in "all other ecosystems existing in the same biome" (Morse et al. 2014). Exclusive to a homogenized ecosystem, which is invaded by invasive species, however, it is very challenging to meet this criterion. The rangelands of eastern Ethiopia for instance Afar rangelands, which are largely invaded by *P. juliflora,* could be considered a novel ecosystem. Although the historical species composition and abundance were not known, it was considered as bare land. Currently, the rangeland is dominantly covered by *P. juliflora* and other herbaceous species that grow under its canopy. Taking the historical plant community, bare land with sparse vegetation, it is possible to say novel ecosystem is created in the Afar rangelands. On the other hand, in the notion of Climate Resilience Green Economy, Ethiopia has performed restoration practices through enrichment plantations for over two decades. Most of the seedlings/saplings distributed have been exotic (Table 8.3). Some of the seedlings/saplings planted that are becoming invasive in several ecosystems are *Acacia saligna* (Labill.) H.L.Wendl. Even though this created an ecosystem with a new combination of species and altered abundance, it is difficult to assess how distinct is in relation to all other ecosystems existing in the same biome. In the absence of quantifiable criteria, judging whether an ecosystem is novel or not becomes ambiguous. Furthermore, the distinction between a hybrid ecosystem, an ecosystem that retained some species from the historical community and added some new species to it, is not also clear.

Table 8.3 Exotic species reported in area exclosures and semi-natural forest ecosystem

No	Species name	Family	Resources	Remark
1	*Acacia decurrens* (J.C.Wendl.) Willd	Fabaceae	Birhane et al. (2017)	Potentially invasive
2	*Acacia saligna* (Labill.) H.L.Wendl	Fabaceae		
3	*Cupressus lusitanica* Mill	Cupressaceae		
4	*Eucalyptus camaldulensis* Dehnh	Myrtaceae		
5	*Eucalyptus globulus* Labill	Myrtaceae		
6	*Grevillea robusta* A.Cunn. ex R.Br	Proteaceae		
7	*Tagetes minuta* L	Asteraceae	Haftay et al. (2013)	
8	*Parthenium hysterophorus* L	Asteraceae		
9	*Senna occidentalis* (L.) Link	Fabaceae		
10	*Casuarina equisetifolia* L	Casuarinaceae	Mengistu et al. (2005)	
11	*Schinus molle* L	Anacardiaceae		
12	*Jacaranda mimosifolia* D.Don	Bignoniaceae	Sinore and Doboch (2021)	
14	*Sesbania sesban* (L.) Merr	Fabaceae		

8.1.4 Thresholds

One of the debating terms in the novel ecosystem is a threshold. Some of the issues raised about thresholds are as follows: (1) Do thresholds exist? (2) Are thresholds easily identifiable and implemented into management? and (3) Are thresholds reversible?

Thresholds, living or non-living, are ecosystem states in which a change in ecosystem properties becomes irreversible, deviating novel ecosystems from historical states (Hobbs et al. 2006; Aronson et al. 2014; Morse et al. 2014). The biotic thresholds include the introduction, deliberate or unintentional, of invasive alien species, which is ideally impossible to eradicate. For example, in Ethiopia, the government and non-governmental organizations invested millions of US dollars each year to eradicate *P. juliflora*. However, the worst scenario is rampantly expanding rather than being contained. This shows that the ecosystem is irreversible to its previous state. Hence, should the decision-makers and policy-makers manage or restore it? This is discussed in the following section. Abiotic thresholds are related to the complex ecosystem process for instance nutrient cycling. (Morse et al. 2014) recommended that both biotic (species composition, species diversity, salinity levels) and abiotic (pH, productivity, decomposition rate, relative change in trophic level biomass, nutrient cycling rates) should be assessed to understand whether the thresholds are crossed or not for a particular ecosystem. It is based on these pre-conditions that decision-makers and managers opt for either restoration or management.

8.2 Shall Novel Ecosystems Be Restored or Managed?

Apart from the controversies emanating from the definitions of a novel ecosystem, should we restore or manage a novel ecosystem is the foci of the debate among scholars. The proponents of a novel ecosystem concept claim that the restoration of novel ecosystems to their historical state is unfeasible (Hobbs et al. 2006, 2009, 2013; Higgs 2017) for ecological and economic reasons. Ecologically, once the novel ecosystems are formed, the threshold is already crossed and it is irreversible. Economically, the fund required to restore novel ecosystems is extremely huge, in which most countries could not afford it. Particularly, developing countries such as sub-Saharan countries are not able to invest millions of US dollars to restore novel ecosystems. Hence, proponents of novel ecosystems and countries in the Global South might favor the management of novel ecosystems and be lured to extract the ecosystem services out of it. Here, the target seems the ultimate ecosystem services they provide to human beings rather than the pristine nature.

On the contrary, opponents of the novel ecosystem concept for instance the hard-core ecologist Simberloff and his colleagues strongly criticized the novel ecosystem concept by saying "ecologist novel ecosystems are a trojan horse for conservation: they provide a license to trash nature if they provide ecosystem services" (Simberloff et al. 2015). Their main concern is that this concept might encourage decision-makers and policy-makers and managers to allow environmentally harmful projects and might deceive the public to exert pressure on the environment. Precisely, this concept could serve as a "Get Out of Jail Free card for companies or individuals trying to avoid investing in research, mitigation or restoration by claiming they are producing novel ecosystems that will provide ecosystem services" (Simberloff et al. 2015). Some scholars argue that there is no need to have the vague "novel ecosystem" concept, which discourages the value of human agency in restoration (Aronson et al. 2014).

8.2.1 *Novel Ecosystem Concept: Conservation Implication in Ethiopia*

I will briefly discuss two types of ecosystems: (1) *P. juliflora* invaded rangelands and (2) Enriched restoration areas.

P. juliflora, native South and Central America, was introduced in the 1970s to limit desertification in eastern Ethiopia. However, it took a few years for all land-use types such as bare lands, croplands, shrublands, grasslands, riparian, and roadsides. Now, nearly 1.17 million hectares are covered by this species only in the Afar rangelands (Shiferaw et al. 2019b). The invasion rate of this invader is 31,127 ha per year causing a loss of $US 602 million ecosystem service values over 31 years (Shiferaw et al. 2019a). Due to the shrinkage of livestock grazing lands, recurrent conflicts are emerging. Consequently, Ethiopia declared the species as a noxious weed in 2017 (NPMS, 2017).

Invasive alien species management is extremely costly. For example, the USA budgets $US 100+ billion annually for invasive alien species management (Pimentel et al. 2005). Furthermore, Canada invests $US 13.3–34.5 billion annually to manage only 16 invasive species (Colautti et al. 2006). Ethiopia is also investing millions of USD annually to manage the invasive *P. juliflora*. However, it is yet to be contained.

Even the pastoralists and agro-pastoralists are willing to pay to manage the invasive species. Studies about the willingness to pay for the management of *P. juliflora* in Afar showed that households are willing to pay $USD 14 (Tilahun et al. 2017) to 50.24 (Bekele et al. 2018). From this perspective, it is reasonable to conclude that let alone restoring to its historical state, it is not even possible to manage the novel ecosystem formed following the introduction of invasive species. This is due to two major reasons. First, developing countries could not afford the fund for such ecosystem restoration. Second, the threshold already crossed beyond reversible state. Hence, I recommend management of such novel ecosystems than restoring it. Nevertheless, maximum effort should be given to the native species composition in that ecosystem.

The other scenario is enriched restoration areas. The ecosystems under this scenario include plantation areas, area exclosure, and enclosures. Unlike the *Prosopis* scenario, restoration of these ecosystems is feasible although empirical data on historical ecosystem states are unavailable. There are plantation areas, secondary forests, area exclosures, and closures, which are enriched through planting exotic species as shown in Table 8.3. The plantation areas and secondary forests are more or less hybrid ecosystems. These ecosystems need to be managed.

The area exclosure and enclosures were first practiced in Tigray, northern Ethiopia, to restore a degraded ecosystem. These experiences were adopted in different parts of the country. The success story is tremendous, particularly in the area exclosures and enclosures in Tigray (Boxes 8.1 and 8.2). In the notion of a novel ecosystem, however, there are looming concerns in these areas. For instance, the enrichment plants, which are exotic and potentially invasive, might overtake the ecosystem, which is dominated by native species. Thus, managing novel ecosystems is more likely to be successful than restoring it in the Ethiopian perspective in particular and Global South in general.

Box 8.1 Successful restoration story from Abreha We Atsbeha, Tigray

Tigray is one of the environmentally degraded areas in Ethiopia. Due to historical reasons, for instance, recurrent war, both local and international, has caused a tremendous impact on the environment. This has led to a scant vegetation cover visited by recurrent droughts. Aligned with the ever-increasing population in the region and transformation of ecosystems, the biodiversity of the region is deteriorating at an alarming.

To halt environmental degradation, Tigray has a long experience of soil and water conservation. This has brought a significant change in vegetation cover after three decades. A study conducted through re-photographing of landscapes and comparing them with the historical landscape that existed 145 years ago revealed that Tigray is currently greener than at any other time in the last 145 years (Nyssen et al. 2014). This is the result of the hard-working people of Tigray and environment-oriented policy.

The Abreha We Atsbeha community in eastern Tigray took an initiative to restore the degraded environment in their district in 1998. Nearly 5000 ha out of 7000 ha were restored through area exclosures in this community (Haile and *Gebregziabher 2020*). This has significantly improved the ecosystem services in the community. The Abreha We Atsbeha community become a model for other restoration projects in the region in particular and outside the region at large.

This was accompanied by regional, national, and international awards and recognitions. For instance, it was awarded a UN Equator Prize at the 2012 RIO+20 summit for their "outstanding projects working to advance sustainable development solutions for people, nature and resilient communities" (https://reliefweb.int/report/ethiopia/ethiopian-village-recognized-rio20-innovative-hunger-solution). It has also made it into documentaries (http://www.tigraycommunitydc.org/press-room/news-articles/the-story-of-abraha-atsbeha/) and blogs (https://jackson.yale.edu/land-restoration-food-security-ethiopia/).

Box 8.2 Eucalyptus plantations in Ethiopia

The exotic fast-growing Eucalyptus was introduced to Ethiopia in the early 1990s to overcome the firewood shortage. Even though the cover of the *Eucalyptus* plantation in the 1970s was about 90, 000 ha, it dramatically increased to about half a million ha in 2006 (FAO 2005). The plantation of Eucalyptus is rampantly increasing and is planted in almost all ecosystem types exclusive to the desert and semi-desert ecosystems in Ethiopia.

However, Eucalyptus is one of the oxymoronic species that is causing ongoing debates both from an ecological (e.g., soil chemical and physical properties, species richness and diversity, eco-hydrology) and economic perspectives. The debate on the importance of Eucalyptus is persisting.

For instance, it is reported that Eucalyptus plantations could improve soil fertility of lands degraded by agriculture although its organic matter is lower than natural vegetation sites (Liang et al. 2016). Its importance in soil erosion prevention, shelterbelts for crops, and soil fertility is positively acknowledged by some authors (Liang et al. 2016). In his review, Zegeye (2010) argues that the positive impacts of Eucalyptus plantations overweigh the negative impacts and the negative impacts emanate from poor management of Eucalyptus plantations.

On the contrary, some researchers claim that *Eucalyptus* is not the right plant to prevent soil erosion and improve soil nutrients (Liang et al. 2016). A comparative study on indigenous and exotic plantation revealed that the species composition is significantly high in the indigenous plantation and indigenous species was not recorded in the exotic species plantation (Wendime et al. 2021). Recent studies argued that *Eucalyptus* plantations constrain regeneration of native woody species and affect species richness (Telila et al. 2015) and weakly support plant species diversity and biotic dispersers such as birds (Lemessa et al. 2021).

References

Aronson J, Murcia C, Kattan GH, et al. (2014) The road to confusion is paved with novel ecosystem labels : a reply to Hobbs et al. Trends Ecol Evol. pp 1–2. https://doi.org/10.1016/j.tree.2014.09.011

Bekele K, Haji J, Legesse B, Schaffner U (2018) Economic impacts of Prosopis spp. invasions on dryland ecosystem services in Ethiopia and Kenya: Evidence from choice. J Arid Environ 158:9–18. https://doi.org/10.1016/j.jaridenv.2018.07.001

Chapin FS, Starfield AM (1997) Time lags and novel ecosystems in response to transient climatic change in arctic Alaska. Clim Chang 35:449–461

Colautti RI, Bailey SA, van Overkijk CDA et al (2006) Characterised and projected costs of nonindigenous species in Canada. Biol Invasions 8:45–59. https://doi.org/10.1007/s10530-005-0236-y

Conradi T, Henriksen MVJ, Svenning JC (2021) Global change, novel ecosystems and the ecological restoration of post-industrial areas: the case of a former brown coal mine in Søby Denmark. Appl Veg Sci 24:e12605. https://doi.org/10.1111/AVSC.12605

Davis AR, Roberts DE, Cummins SP (1997) Rapid invasion of a sponge-dominated deep-reef by Caulerpa scalpelliformis (Chlorophyta) in Botany Bay, New South Wales. Aust J Ecol 22:146–150. https://doi.org/10.1111/J.1442-9993.1997.TB00653.X

Ellis EC, Goldewijk KK, Siebert S, Lightman D (2010) Anthropogenic transformation of the biomes, 1700 to 2000. Glob Ecol Biogeogr 19:589–606. https://doi.org/10.1111/j.1466-8238.2010.00540.x

FAO (2005) Global Forest Resources Assessment. FAO Forestry Paper 140, Rome, Italy

Fehr V, Buitenwerf R, Svenning J-C (2020) Non-native palms (Arecaceae) as generators of novel ecosystems: a global assessment. Divers Distrib 26:1523–1538. https://doi.org/10.1111/ddi.13150

Ferraz S, Brancalion PHS, Meli P (2020) On the need to differentiate the temporal trajectories of ecosystem structure and functions in restoration programs. Trop Conserv Sci 13:1–6. https://doi.org/10.1177/1940082920910314

Haftay H, Yayneshet T, Animut G, Treydte AC (2013) Rangeland vegetation responses to traditional enclosure management in eastern Ethiopia. Rangel J 35:29–36. https://doi.org/10.1071/RJ12054

Higgs E (2017) Novel and designed ecosystems. Restor Ecol 25:8–13. https://doi.org/10.1111/rec.12410

Hobbs RJ, Arico S, Aronson J et al (2006) Novel ecosystems: theoretical and management aspects of the new ecological world order. Glob Ecol Biogeogr 15:1–7. https://doi.org/10.1111/j.1466-822x.2006.00212.x

Hobbs RJ, Higgs E, Harris JA (2009) Novel ecosystems: implications for conservation and restoration. Trends Ecol Evol 24:599–605. https://doi.org/10.1016/j.tree.2009.05.012

Hobbs RJ, Higgs ES, Hall CM (2013) Novel ecosystems: intervening in the new ecological world order. Wiley, UK

Lemessa D, Mewded B, Legesse A et al (2021) Do ecological evidences support biodiversity optimizing roles of Eucalyptus plantation forests? SSRN Electron J. https://doi.org/10.2139/ssrn.3954989

Levin SA (1992) The problem of pattern and scale in ecology. Ecology 73:1943–1967. https://doi.org/10.2307/1941447

Lewis CL, Granek EF (2021) Drivers of zooplankton community composition in a novel ecosystem: Hawai'i mangroves as a case study. J Appl Ecol 58:3023–3035. https://doi.org/10.1111/1365-2664.14028

Liang J, Reynolds T, Wassie A et al (2016) Effects of exotic Eucalyptus spp. plantations on soil properties in and around sacred natural sites in the northern Ethiopian Highlands. AIMS Agric Food 1:175–193. https://doi.org/10.3934/agrfood.2016.2.175

Lugo AE, Helmer E (2004) Emerging forests on abandoned land: Puerto Rico's new forests. For Ecol Manage 190:145–161. https://doi.org/10.1016/j.foreco.2003.09.012

Mascaro J, Hughes RF, Schnitzer SA (2012) Novel forests maintain ecosystem processes after the decline of native tree species. Ecol Monogr 82:221–228. https://doi.org/10.1890/11-1014.1

McCann JC (2017) The history of agriculture in Ethiopia. Oxford Res Encycl Afr Hist. pp 1–14

Mengistu T, Teketay D, Hulten H, Yemshaw Y (2005) The role of enclosures in the recovery of woody vegetation in degraded dryland hillsides of central and northern Ethiopia. J Arid Environ 60:259–281. https://doi.org/10.1016/j.jaridenv.2004.03.014

Morse NB, Pellissier PA, Cianciola EN, et al. (2014) pragmatic applications Novel ecosystems in the Anthropocene : a revision of the novel ecosystem. Ecol Soc 19:12. https://doi.org/10.5751/ES-06192-190212 Insight

Murcia C, Aronson J, Kattan GH et al (2014) A critique of the 'novel ecosystem' concept. Trends Ecol Evol 29:548–553. https://doi.org/10.1016/j.tree.2014.07.006

Norris CE, Dungait JAJ, Joynes A, Quideau SA (2013) Organic Geochemistry Biomarkers of novel ecosystem development in boreal forest soils. Org Geochem 64:9–18. https://doi.org/10.1016/j.orggeochem.2013.08.014

Orgensen JG, BRenner SJ., GReenwalt LR, Vrtiska MP (2021) Decline of novel ecosystems used by endangered species : the case of piping plovers, least terns, and aggregate mines. Ecosphere 12:e03474. https://doi.org/10.1002/ecs2.3474

Pimentel D, Zuniga R, Morrison D (2005) Update on the environmental and economic costs associated with alien invasive species in the United States. Ecol Econ 52:273–288

Seastedt TR, Hobbs RJ, Suding KN (2008) Management of novel ecosystems: are novel approaches required? Front Ecol Environ 6:547–553. https://doi.org/10.1890/070046

Shiferaw H, Bewket W, Alamirew T et al (2019a) Implications of land use/land cover dynamics and Prosopis invasion on ecosystem service values in Afar Region, Ethiopia. Sci Total Environ Science 67:354–436. https://doi.org/10.1016/j.scitotenv.2019.04.220

Shiferaw H, Scha U, Bew W, et al. (2019b) Modelling the current fractional cover of an invasive alien plant and drivers of its invasion in a dryland ecosystem. pp 1–12. https://doi.org/10.1038/s41598-018-36587-7

Simberloff D, Murcia C, Aronson J (2015) OPINION: "Novel ecosystems" are a Trojan horse for conservation: they provide a license to trash nature if they provide ecosystem services. https://ensia.com/voices/novel-ecosystems-are-a-trojan-horse-for-con. Accessed 25 May 2021

Sinore T, Doboch D (2021) Area exclosure effects on soil properties and woody species diversity in Amelake sub-watershed, Southern Ethiopia. Int J River Basin Manag. https://doi.org/10.1080/15715124.2020.1843469

Solórzano A, Brasil-machado A, de Oliveira R (2021) Land use and social-ecological legacies of Rio de Janeiro's Atlantic urban forests: from charcoal production to novel ecosystems. R Soc Open Sci 8:201855. https://doi.org/10.1098/rsos.201855

Strayer DL (2010) Alien species in fresh waters: ecological effects, interactions with other stressors, and prospects for the future. Freshw Biol 55:152–174. https://doi.org/10.1111/j.1365-2427.2009.02380.x

Telila H, Hylander K, Nemomissa S (2015) The potential of small Eucalyptus plantations in farmscapes to foster native woody plant diversity: local and landscape constraints. Restor Ecol 23:918–926. https://doi.org/10.1111/rec.12257

Tilahun M, Birner R, Ilukor J (2017) Household-level preferences for mitigation of Prosopis juliflora invasion in the Afar region of Ethiopia: a contingent valuation. J Environ Plan Manag 60:282–308. https://doi.org/10.1080/09640568.2016.1152955

Wendime H, Lemessa D, Bekele D (2021) A comparative analysis of indigenous and exotic tree species management practices in agricultural landscapes of Southwest Ethiopia. Trees for People 4:100059. https://doi.org/10.1016/j.tfp.2020.100059

Zegeye H (2010) Environmental and socio-economic implications of Eucalyptus Ethiopia. In: Gil L, Tadesse W, Tolosana E, López R (eds) Proceedings of the conference on eucalyptus species management, history, status and trends in Ethiopia. Ethiopian Institute of Agricultural Research, Addis Ababa, Ethiopia, pp 184–205

Chapter 9
Ecological Connectivity

Abstract Ecological connectivity is the degree to which the landscape impedes or facilitates movement of organisms, i.e., genes, propagules, individuals, and populations, and combines both spatial and functional features. The human-dominated landscape with patched and isolated natural vegetation such as forests or grasslands should consider ecological connectivity. Buffer zones, corridors, and stepping stones are important components of ecological connectivity. Home gardens, farmland, Church forests, and roadsides need to be considered in ecological connectivity.

Keywords Connectivity · Neophytes · Sacred forests · Stepping stones · Tree on farm

9.1 Introduction

Fragmentation and connectivity are common features of a landscape (Jongman 2019). Several countries such as Ethiopia have been applying the traditional nature conservation approach, which is often implemented by establishing protected areas (core area only). The traditional conservation approach is simply establishing a protected area giving less emphasis to ecological networks such as corridors and connectivity. However, man has shaped the natural ecosystem consequently large and intact natural ecosystems are sparse. Contemporary conservation approaches on the other hand recognize the contribution of habitat connectivity to the conservation of biodiversity. However, what is ecological/habitat connectivity?

Ecological connectivity is defined as the degree to which the landscape impedes or facilitates movement of organisms, i.e., genes, propagules, individuals, and populations (Taylor et al. 1993; Rudnick et al. 2012) that combines both structural and functional features (With et al. 1997). Structural feature is the spatial relationships such as intactness between the structural elements of the landscape, which is independent of the ecological characteristics of the species, whereas functional connectivity is a landscape characteristic related to the ecology of the species such as the movement of species between habitat patches (Taylor et al. 1993). Functional characteristics of connectivity are strongly associated with the metapopulation or metacommunity theory (Chap. 2). If the structural and functional connectivity is ensured

K. G. Yaynemsa, *Plant Biodiversity Conservation in Ethiopia*,
https://doi.org/10.1007/978-3-031-20225-4_9

in a landscape, it is more likely that the ecosystem functions and biodiversity are maintained. For example, connectivity increases plant–pollinator interactions in a landscape (Chap. 6). This in turn increases productivity. In their review, Fletcher Jr et al. (2016) revealed consistent positive effects of connectivity from genes to communities. Connectivity comprises three major components, viz. core area, buffer zone, and ecological corridors (Jongman and Pungetti 2004) (Fig. 9.1). Core areas are properly managed protected areas with high natural values, areas of conservation of habitats, species, or landscapes. Buffer zone is a spatial zone of a protected areas that minimizes human impact on biodiversity around the core areas, and it halts ecological isolation. On the other hand, ecological corridor is a physical landscape structure that is designed to enhance ecological connectivity.

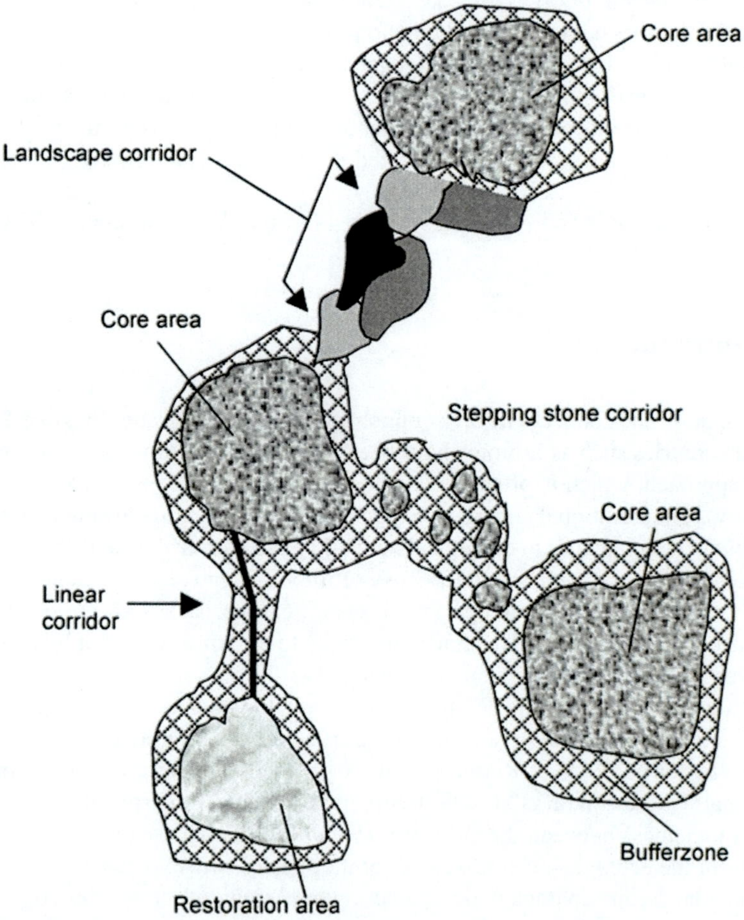

Fig. 9.1 Demonstration of ecological networks (Bouwma et al. 2002)

The human-dominated landscape with patched and isolated natural vegetation such as forests or grasslands should consider ecological connectivity. The protected areas in Ethiopia are isolated and unconnected. Buffer zones, corridors, and stepping stones are important components of ecological connectivity. However, these extremely important components are missing in protected areas in Ethiopia. Exclusive to protected areas, small patches of vegetation that play an important role in plant dispersal and animal movements are not getting due attention. The ecosystem and biodiversity crisis would be worst unless and otherwise the conservation approach is revisited. The Convention on Biodiversity recognized the value of habitat connectivity under Aichi Target 11 to be reached in 2020. Though Ethiopia is a signatory to CBD and claimed that it achieved the Aichi Target 11 (Bacon et al. 2019), connectivity is poorly implemented both in the protected areas and outside of the protected area. Probably, the absence of ratified land-use policy may have its own impact on the ineffectiveness of connectivity in protected areas. The already established protected area networks may not benefit a lot from the land-use policy to be ratified in the future. However, the policy would have a vital contribution to the proposed protected areas and natural reserves to be established in the future.

There is accumulated literature on land use and land cover change in Ethiopia. However, neither of these studies instigated connectivity albeit scanty articles on the impact of corridors on particular animal species. Recently, a study at county level explained the importance of using Church forests and area closures (Demissie et al. 2022). Landscape matrices analyses are also scanty at the national level. This might have its share for the poor implementation of ecological/habitat connectivity inside/outside protected areas. Hence, future studies and researchers should emphasize on connectivity and restoration, connectivity and climate change, which method is appropriate to design connectivity so that help policy-makers and decision-makers.

Designing ecological connectivity in anthropic landscapes might not be an easy task due to the myriad variables to be considered. I am recommending the variables to be considered in designing connectivity in the future. The following sub-sections discussed the variables, i.e., home gardens, farmlands, Church forests, and roadsides that are important for ecological connectivity in human-dominated landscapes (Fig. 9.5).

9.1.1 Home Gardens

Biodiversity is known to exist everywhere including a natural dense forest to barren land in a terrestrial ecosystem and marine and freshwater in an aquatic ecosystem. It is obvious that agricultural expansion and settlements cause a massive impact on biodiversity. Both of them are the primary drivers of habitat destruction/loss in most terrestrial biomes. Besides the detrimental impact of agricultural expansion and settlements, they are also a habitat for a considerable number of plant species. Home gardens are important places for biodiversity. Some studies even claimed that home gardens are repositories for some life forms such as epiphytes (Hylander and

Nemomissa 2008). Other studies also implied the value of home gardens as a source of herbal medicine, spice, fragrance, and ornamental plants. Their contribution in mitigating climate change by stocking soil organic carbon was also discussed in Birhane et al. (2020).

Forty-nine woody species with several indigenous tree species including *Juniperus procera* Hochst. ex Endl., *Hagenia abyssinica* (Bruce) J.F.Gmel., *Cordia africana* Lam., and *Croton macrostachyus* Hochst. ex Delile with a Shannon diversity of 1.87 were reported in southern Ethiopia (Birhane et al. 2020). Of the 56 species recorded in the Assosa district, 31 (55.35%) species are indigenous (Tadesse et al. 2019). These include species such as *C. africana, Syzygium guineense* (Willd.) DC., *Terminalia brownii* Fresen., *C. macrostachyus, Entada abyssinica* Steud. ex A.Rich., *Ficus sur* Forssk, *Albizia gummifera* (J. F. Gmel.) C. A. Sm., and *Pilostigma thonningii* (Schumach.) Milne-Redh. *S. guineense* is one of the species, which is over-exploited from the natural forests and protected areas. In their comparative study, Haile et al. (2016) revealed home gardens hosted high woody species than grazing land, Eucalyptus woodlots, and farmland in Meskan district, central Ethiopia. The indigenous plant species richness in home gardens is also higher than the domesticated plant species. *C. africana, C. macrostachyus, Acacia albida* Delile., and *Acacia abyssinica* Hochst. ex Benth. are some of the indigenous vascular plants reported to occur in home gardens. This proves home gardens play a vital role in conserving indigenous and threatened plants. Although a considerable number of indigenous species are hosted in the home gardens, there are also numerous exotic species. Working on home gardens would have a great impact on enhancing ecological connectivity. It is not only plant species that exist in home gardens, other taxa including pollinators are also common.

A study in Jimma, Ethiopia, recorded higher bird species, pollinators, and/or dispersers, in home gardens than Moist evergreen Afromontane Forest (Engelen et al. 2017). This is an interesting finding. Obviously, birds are both long-distance dispersal agents and pollinators as well. Consequently, these species play a fundamental role in connecting the ecological networks between the natural forests, woodlands, home gardens, farmlands, and other land-use types. As indicated earlier, home gardens in particular and other land-use types at large must be considered in designing ecological/habitat connectivity.

The ever-increasing population globally is putting pressure on biodiversity. As a result, habitat transformations are happening almost in every terrestrial biome of the world. Hence, home gardens might be a good asset for conservation. In some cases, these might be the only option to conserve threatened plant species. For example, Tsegaye et al. (2010) claimed that home gardens are the best option to sustain the declining population of *Dobera glabra* (Forssk.) Juss. ex Poir. and improve its recruitment. Moreover, home gardens can serve as *refugia* to medicinal plants from drought and climate change (Teklehaymanot 2017).

Generally, the importance of home gardens is multifaceted. Taking their value in biodiversity conservation and ecological connectivity into account, home gardens could be part of the ecological networks of the protected areas. Nevertheless, it demands a tremendous effort to aware the people, decision-makers, and policy-makers. Even though most of the studies in home gardens reported indigenous plants are dominant, in some cases almost equal numbers of exotic species were reported. Thus, aware people what species to grow in their homesteads, discussing why is it important for biodiversity conservation, and reaching into consensus on why they should do that must be a prior agenda because integrating home gardens in ecological connectivity might be challenging taking into account that home gardens are individual properties.

Despite the ecological importance, vegetation ecology studies in urban landscapes of Ethiopia are scanty. However, a study by Fetene and Worku (2013) documented 44 woody species in urban forestry in Addis Ababa, Ethiopia. Some endemic plants such as *Erythrina brucei* Schweinf. were documented in the field survey. Since urban ecology is an important component of ecological connectivity, it should not be ignored in biodiversity conservation. Green infrastructure (Jongman 2019) is an efficient way of maintaining functional connectivity in an urban area (Xun et al. 2014) and the nearby agricultural landscapes, home gardens, and Church forests.

9.1.2 Farmland

Human beings shape most of the ecosystem transformations in the Anthropocene. Agricultural/cropland expansion at the expense of natural vegetation is a common phenomenon globally and is severe in developing countries such as Ethiopia. Reconstruction of pre-agricultural land expansion vegetation cover of Ethiopia by Hailu et al. (2015) showed that more than one-third of today's farmlands were previously covered mainly by broad-leaved evergreen and deciduous forest. When such ecologically important ecosystems are transformed into agricultural land or land use that barely supports biodiversity, the biodiversity within these ecosystems is probably extinct or threatened. Therefore, to mitigate the extinction of tropical forest biodiversity, it is becoming increasingly important to focus on the role of agricultural landscapes in preserving biodiversity (Hylander and Nemomissa 2008). Conservation of isolated tree species in an agricultural landscape (tree on farm) improves the likelihood of achieving the conservation objectives of comprehensiveness, adequacy, and representativeness (Gibbons and Boak 2002). The agricultural landscape could significantly preserve a large number of species (Fig. 9.2). As a result, their value in ecological networks and connectivity could not be neglected because they play a crucial role in maintaining biodiversity in human-dominated landscapes. Some authors have recommended designing biodiversity-based ecosystem services in the agricultural landscape (Landis 2017; Cadavid-florez et al. 2020).

Tadesse et al. (2019) recorded 22 vascular plant species in parklands. Of these, eight are exotic and the remaining are indigenous. *S. guineense, C. africana, T. brownii, F. sur, Combretum aculeatum* Vent., and *Allophylus abyssinicus* (Hochst.) Radlk. are some of the indigenous woody species recorded in Assosa district, western Ethiopia. Haile et al. (2016) also reported 25 woody species in central Ethiopia

Fig. 9.2 Photograph of tree species inside agricultural landscape in Abijatta-Shalla Lakes National Park: © Kflay Gebrehiwot

though exotic species are dominant over the indigenous woody plants. Furthermore, 42 woody species such as *A. abyssinica* subsp. *abyssinica, C. macrostachyus, Anogeissus leiocarpus* A.Rich., *Balanites aegyptiaca* (L.) Delile, *Pittosporum viridiflorum* Sims, and *Pterocarpus lucens* Lepr. ex Guill. and Perr. were reported in the agricultural landscape in Tigray, Ethiopia (Gebrewahid and Abrehe 2019).

The threatened species, *Boswellia papyrifera* (Del.) Hochst, is also reported to have been found in this agricultural landscape. A study in Eastern Shewa (Ethiopia) also reported 77 woody species, of which 70% are indigenous (Endale et al. 2017). *Acacia nilotica* (L.) Willd. ex Delile, *A. gummifera, B. aegyptiaca, Erythrina abyssinica* Lam., *A. albida*, and *Ficus sycomorus* L. are some of the indigenous species documented. A coffee landscape in southwestern Ethiopia is *refugia* for native, endemic, and threatened species (Tadesse et al. 2014). The threatened species include *Baphia abyssinica* Brummitt, *Milicia excelsa* (Welw.) C.C.Berg, *Ocotea kenyensis* (Chiov.) Robyns and R.Wilczek, *Prunus africana* (Hook.f.) Kalkman, and *Rinorea friisii* M.G.Gilbert. The endemic species include *Bothriocline schimperi* Oliv. and Hiern ex Benth., *E. brucei, Euphorbia dumalis* S. Carter, and *Vepris dainellii* (Pichi-Serm.) Kokwaro. The existence of threatened and endemic species in agricultural landscapes is an excellent implication of why they should be considered in ecological connectivity and networks.

Even though isolated tree species in agricultural landscape provides multifaceted values, their role in landscape connectivity is undervalued. For example, small wooded patches and isolated trees in the agricultural landscape contributed from 26 to 59% to overall connectivity (Cadavid-florez et al. 2020). This includes living trees serving as fences and isolated trees. If agricultural landscapes are designed in a way that benefits biodiversity and ecosystem services, plant–pollinator interactions are also more sustained. Consequently, the maximum ecological connectivity could be achieved between the previously discussed, home gardens, agricultural land, protected areas, and vice versa.

9.1.3 Church Forests

Religious institutions have endured contribution to forest conservation in Ethiopia. Ethiopian Orthodox Tewahedo Church (EOTC) (Fig. 9.3) is in the front-line in this regard. In the human-dominated landscapes of northern Ethiopia, the Church forests are conspicuous. It is due to the trees around the Church unlike their counterpart agricultural land or barren lands. Apart from the countless ecosystem services provided, EOTC forests are a habitat for numerous species of different taxa. Hosting many rare and endemic species within it, such forests are *refugia* of biodiversity (Dudley et al. 2009; Gebrehiwot et al. 2019, 2020) that are disappeared in most parts of northern Ethiopia (Linder 2005; Wassie et al. 2005). In addition, Church forests are considered as the last natural seed banks for native trees species, reference areas for local endemism, last cornerstones for species distribution, seed source for rehabilitation for degraded areas, and important reserves for species and biodiversity conservation and buffer against depletion (Kassahun and Bender 2019).

A study in south Gonder, Ethiopia, documented 97 species, of which five are exotic in eight Church forests with 1.6–100 ha size (mean = 20.23 ha) (Wassie et al. 2005). Their social survey also implied people residing around these Church forests are also well aware of the importance of these forests and the contribution of EOTC. A survey on 78 Church forests in northern Ethiopia documented 148 indigenous woody species which comprise >50% of the tree species present in the flora of tropical northeast Africa although individual forests were small (2.48 ± 0.24 ha) and isolated (Aerts et al. 2016). Four plant communities were identified in the surveyed forests. These are *Olea-Acacia, Juniperus-Olea, Juniperus-Podocarpus,* and *Juniperus.* Considerably, high species richness and endemic vascular plants are also reported in the human-dominated landscape of the Abune Yosef mountain range (Gebrehiwot et al. 2019). Apart from vascular plants, LoTemplio et al. (2016) documented 3039 macro-invertebrates in south Gondar zonal administrative State Church forests. Despite the significant increase in forest size (8.2%, 0.41 ha) over a 50 year time series, a significant decrease (16.9%) in crown cover over the same time series

Fig. 9.3 A Google Earth image of Yimrehane Kiristos Church forest in an extremely degraded landscape

was documented. This is followed by significant human disturbances such as exotic plantations and trails (Cardelus et al. 2017; Kassahun and Bender 2019). The disturbance was high across all forests (56%) and was negatively associated with tree species richness, density, and biomass, and seedling richness and density (Cardelus et al. 2019). However, the underlying threat to sustaining Church forests is a lack of clear boundary demarcation (Wassie et al. 2005). Long-term conservation of biodiversity of individual patches and evolutionary potential of species may be threatened by isolation, small sizes of tree species populations, and disturbance, especially when considering climate change (Aerts et al. 2016).

Even though the Church forests are isolated and small in size, seed dispersal is not disrupted in wild olive trees and that large protected populations are very important to conserve genetic resources. However, the higher level of inbreeding and evidence for population bottlenecks in the small populations, as well as the persisting heavy pressure on most remaining populations, warrants quick action to maintain the genetic diversity of wild olive in the Ethiopian highlands (Kassahun and Bender 2019). Hence, interconnecting the isolated and small patches by vegetation corridors, creating buffering areas and livestock fencing, and reforesting are suitable measures (Kassahun and Bender 2019).

Despite their value has been considered safe haven for tree [indigenous and endemic] species, the Church forests are under eminent threat. A survey of Church forest network around Lake Tana, Ethiopia, warned that the Church forests surveyed face strong extinction debts, with five tree species at risk of completely disappearing from the region and many more species only regenerating in a limited number of Church forests (Mequanint et al. 2020).

Thus, clear demarcation of the Church forests, legal protection in collaboration with the EOTC, minimizing exotic plantations, grazing, and encroachment should be taken into account to solve the threats. As a result, Church forests would be the source of indigenous propagules for nearby land uses so that it enhances ecological connectivity provided that a suitable environment is set in the surrounding farmlands and home gardens. The flow of genetic resources from the farmlands and home gardens to the Church forests could also be improved and maintained. Long-term ecological connectivity among home gardens, farmlands, Church forests, and protected areas could be maintained.

9.1.4 Roadsides

Road networks are important infrastructures that connect administrative boundaries such as urban and rural areas at country and cross-country levels. As a matter of this fact, Ethiopia had about 144,024 km at the end of 2017/18 fiscal year (Gebrehiwot 2021). Despite their importance in development, road networks in most parts of Ethiopia are facilitating soil erosion and gully formations, and they are also the main conduit for invasive, ruderal, and neophytes (Fig. 9.4).

Roadside networks as a channel for the introduction of invasive, ruderal, and neophytes are well documented globally (Mcdougall et al. 2018; Bashir and Ahmad 2019; Gonzalez et al. 2021; Pérez-postigo et al. 2021; Turner et al. 2021). In Ethiopia, however, this topic is ignored by the academia. Literature search on Scopus and Google Scholar on 20 December 2021 showed studies on this theme are scanty, if not non-extant.

As stated above, roadside networks are ecologically important for two reasons. First, they serve as a conduit to the introduction of invasive, ruderals, and neophytes. Once, these species are introduced through the roadside networks, it is likely that they will invade the landscape and modify the ecosystem. Second, roadside networks, if they are managed poorly, they aggravate soil erosion by forming a gully and flood. It is obvious that soil erosion and degradations negatively affect plant biodiversity in particular and ecosystem at large. Even, in some rural areas in Tigray 10–50% of yield reduction was reported because of the dust coming from the road networks (Perez et al. 2016). Nevertheless, roadside ecology did not receive the deserved attention. Hence, researchers, conservation practitioners, policy-makers, and decision-makers might need to gear their attention to work on this theme.

Furthermore, practically it is possible to consider roadside networks to connect different habitats for instance farmland, home gardens, Church forests, natural forests, and others and promote ecologically sustainable development. This could be effectively implemented by planting native species along the roadside gradient. The native species to be planted could be small to medium trees, shrubs, or grasses. According to the competition theory, the native plants would probably [more likely if the functional traits are more related] have a vital role in managing the invasive species. The mass plantation (Chaps. 7 and 8) reported in Ethiopia annually did not

Fig. 9.4 Ecological importance of roadsides. Roads facilitating soil erosion by gully formation (**a** and **b**) and roads as an important channel for the introduction of ruderal, invasive, and neophytes (**c**). At least three invasive species, namely *Lantana camara* L., *Calotropis procera* (Aiton) W.T. Aiton, and *Nicotiana glauca* Graham, were observed in **c**

consider the roadside networks; rather, the focus has been on either degraded forests or bare lands with sparse vegetation. Hence, plantation projects either in mass or small projects need to consider roadside networks in future restoration practices.

Monitoring and evaluating the effectiveness of the roadside network plantation must be performed regularly. Furthermore, it is advisable to integrate experts from several disciplines such as botany, ecology, soil, and geology while implementing the design (Fig. 9.5).

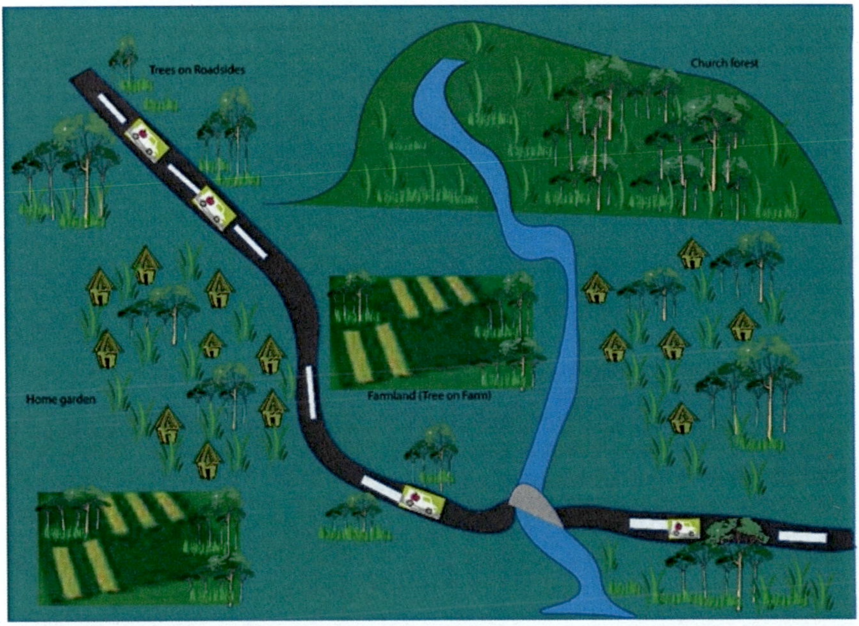

Fig. 9.5 A simple demonstration of ecological connectivity

References

Aerts R, Van Overtveld K, November E et al (2016) Conservation of the Ethiopian church forests: threats, opportunities and implications for their management. Sci Total Environ 551–552:404–414. https://doi.org/10.1016/j.scitotenv.2016.02.034

Bacon E, Gannon P, Stephen S et al (2019) Aichi Biodiversity Target 11 in the like-minded megadiverse countries. J Nat Conserv 51:125723. https://doi.org/10.1016/j.jnc.2019.125723

Bashir H, Ahmad SS (2019) Ordination classification of relationship of vegetation and soil's edaphic factors along the roadsides of Wahcantt, Pakistan. J King Saud Univ—Sci 31:511–518. https://doi.org/10.1016/j.jksus.2018.03.006

Birhane E, Ahmed S, Hailemariam M, Negash M (2020) Carbon stock and woody species diversity in homegarden agroforestry along an elevation gradient in southern. Agrofor Syst 4. https://doi.org/10.1007/s10457-019-00475-4

Bouwma IM, Jongman RHG, Butovsky RO (2002) Indicative map of the Pan-Document., European ecological network for central and Eastern Europe. Technical background. Tilburg/Budapest

Cadavid-florez L, Laborde J, James D et al (2020) Isolated trees and small woody patches greatly contribute to connectivity in highly fragmented tropical landscapes. Landsc Urban Plan 196:103745. https://doi.org/10.1016/j.landurbplan.2020.103745

Cardelus CL, Scull P, Eshete AW, et al. (2017) Shadow conservation and the persistence of sacred church forests in northern. Biotropica 0:1–8. https://doi.org/10.1111/btp.12431

Cardelus CL, Woods CLB, Mekonnen A et al (2019) Human disturbance impacts the integrity of sacred church forests Ethiopia. PLoS ONE 14:e0212430. https://doi.org/10.1371/journal.pone.0212430

Demissie F, Yeshitela K, Kindu M (2022) Ecological status and plan for connectivity of fragmented forests as a means of degraded land restoration in South Gonder, Ethiopia. In: Kindu M, Schneider

T, Wassie A, et al. (eds) State ofthe art in Ethiopian Church Forests and restoration options. Springer Nature, Switzerland AG

Dudley N, Higgins-Zogib L, Mansourian S (2009) The links between protected areas, faiths, and sacred natural sites. Conserv Biol 23:568–577

Endale Y, Derero A, Argaw M, Muthuri C (2017) Farmland tree species diversity and spatial distribution pattern in semi-arid East Shewa, Ethiopia. For Trees Livelihoods 8028:1–16. https://doi.org/10.1080/14728028.2016.1266971

Engelen D, Lemessa D, Şekercioğlu ÇH, Hylander K (2017) Similar bird communities in home-gardens at different distances from Afromontane forests. Bird Conserv Int 27:83–95. https://doi.org/10.1017/S0959270916000162

Fetene A, Worku H (2013) Planning for the conservation and sustainable use of urban forestry in Addis Ababa, Ethiopia. Urban for Urban Green 12:367–379

Fletcher RJ Jr, Burrell NS, Reichert BE et al (2016) Divergent perspectives on landscape connectivity reveal consistent effects from genes to communities. Curr Landsc Ecol Rep 1:67–79. https://doi.org/10.1007/s40823-016-0009-6

Gebrehiwot F (2021) Ethiopia—roads, railways and logistics. https://www.trade.gov/country-commercial-guides/ethiopia-roads-railways-and-logistics. Accessed 20 Dec 2021

Gebrehiwot K, Demissew S, Woldu Z et al (2019) Elevational changes in vascular plants richness, diversity, and distribution pattern in Abune Yosef mountain range, Northern Ethiopia. Plant Divers 41:220–228. https://doi.org/10.1016/j.pld.2019.06.005

Gebrehiwot K, Woldu Z, Fekadu M et al (2020) Classification and ordination of plant communities in Abune Yosef mountain range, northern Ethiopia. Acta Ecol Sin 40:398–411. https://doi.org/10.1016/j.chnaes.2019.12.001

Gebrewahid Y, Abrehe S (2019) Biodiversity conservation through indigenous agricultural practices: Woody species composition, density and diversity along an altitudinal gradient of Northern Ethiopia. Cogent Food Agric 5:1–30. https://doi.org/10.1080/23311932.2019.1700744

Gibbons BP, Boak M (2002) The value of paddock trees for regional conservation in an agricultural landscape. Ecol Manag Restor 3:205–210

Gonzalez L, Rovere AE, Ailin G, Id C (2021) Alien plant species on roadsides of the northwestern Patagonian steppe (Argentina). PLoS ONE 16:e0246657. https://doi.org/10.1371/journal.pone.0246657

Haile G, Lemenih M, Senbeta F, Itanna F (2016) Plant diversity and determinant factors across smallholder agricultural management units in Central Ethiopia. Agrofor Syst. https://doi.org/10.1007/s10457-016-0038-5

Hailu B, Eiji E, Heiskanen J, Pellikka P (2015) Reconstructing pre-agricultural expansion vegetation cover of Ethiopia. Appl Geogr 62:357–365. https://doi.org/10.1016/j.apgeog.2015.05.013

Hylander K, Nemomissa S (2008) Home garden coffee as a repository of epiphyte biodiversity in Ethiopia. Front Ecol Environ 6:524–528. https://doi.org/10.1890/080001

Jongman RHG (2019) Encyclopedia of ecology. Connectivity and Ecological Networks, pp 366–376

Jongman RHG, Pungetti GP (eds) (2004) Ecological networks and greenways, concepts methods and implementation. Cambridge University Press, Cambridge, UK

Kassahun T, Bender S (2019) Saving the last endemic-church forests in Ethiopia: the case of Lake tana biosphere reserve. In: Filho WL, Barbir J, Preziosi R (eds) Handbook of climate change and biodiversity, climate change management. Springer Nature, Switzerland, AG

Landis DA (2017) Designing agricultural landscapes for biodiversity-based ecosystem services. Basic Appl Ecol 18:1–12. https://doi.org/10.1016/j.baae.2016.07.005

Linder HP (2005) Evolution of diversity: the Cape flora. p 10. https://doi.org/10.1016/j.tplants.2005.09.006

LoTemplio S, Reynolds TW, Eshete AW et al (2016) Ethiopian Orthodox church forests provide regulating and habitat services: evidence from stream sediment and aquatic insect analyses. Afr J Ecol 55:247–251

Mcdougall KL, Lembrechts J, Rew LJ, et al. (2018) Running off the road: roadside non-native plants invading mountain vegetation. 0123456789:3461–3473. https://doi.org/10.1007/s10530-018-1787-z

Mequanint F, Wassie A, Aynalem S, Strubbe D (2020) Biodiversity conservation in the sacred groves of north-west Ethiopia: diversity and community structure of woody species. Glob Ecol Conserv 24:e01377. https://doi.org/10.1016/j.gecco.2020.e01377

Pérez-postigo I, Bendix J, Vibrans H, Cuevas-guzmán R (2021) Diversity of alien roadside herbs along an elevational gradient in western Mexico. NeoBiota 65:71–91. https://doi.org/10.3897/neobiota.65.67192

Perez MA, Tegebu FN, Van SF (2016) Roadside planting in Ethiopia: turning a problem into an opportunity. Sustain Environ 1:98–115

Rudnick DA, Ryan SJ, Beier P, et al. (2012) Published by the Ecological Society of America esa

Tadesse E, Abdulkedir A, Khamzina A et al (2019) Contrasting species diversity and values in home gardens and traditional parkland agroforestry systems in Ethiopian sub-humid lowlands. Forests 10:266. https://doi.org/10.3390/f10030266

Tadesse G, Zavaleta E, Shennan C (2014) Coffee landscapes as refugia for native woody biodiversity as forest loss continues in southwest Ethiopia. Biol Conserv 169:384–391. https://doi.org/10.1016/j.biocon.2013.11.034

Taylor PD, Fahrig L, Henein K, Merriam G (1993) Connectivity is a vital element of landscape structure. Oikos 68:571–573

Teklehaymanot T (2017) An ethnobotanical survey of medicinal and edible plants of Yalo Woreda in Afar regional state, an ethnobotanical survey of medicinal and edible plants of Yalo Woreda in Afar regional state Ethiopia. J Ethnobiol Ethnomed 13:40. https://doi.org/10.1186/s13002-017-0166-7

Tsegaye D, Haile M, Moe SR (2010) The effect of land use on the recruitment and population structure of the important food and fodder plant, Dobera glabra (Forssk.) Poir., in northern. J Arid Environ 74:1074–1082. https://doi.org/10.1016/j.jaridenv.2010.03.003

Turner SC, Esler KJ, Kalwij JM (2021) Road verges facilitate exotic species' expansion into undisturbed natural montane grasslands. Appl Veg Sci 24:1–9. https://doi.org/10.1111/avsc.12615

Wassie A, Teketay D, Powell N (2005) Church forests in north Gonder administrative zone, northern Ethiopia. For Trees Livelihoods ISSN 15:349–373. https://doi.org/10.1080/14728028.2005.9752536

With KA, Gardner RH, Turner MG (1997) Landscape connectivity and population distributions in heterogeneous environments. Oikos 78:151–169

Xun B, Yu D, Liu Y (2014) Habitat connectivity analysis for conservation implications in an urban area. Acta Ecol Sin 34:44–52. https://doi.org/10.1016/j.chnaes.2013.11.006

Chapter 10
Strategic Conservation Planning Approach

Abstract Understanding the critical constraints mainly ecological, social, and financial constraints would have a critical impact on the success of a proposed conservation project. Although a significant number of species are on the brink of extinction, the budget required to conserve all threatened species is limited globally. This could have a profound influence on the planning methods and conservation strategies. Hence, conservation planning and priorities should consider the aforementioned constraints to achieve the set conservation goals. Evidence-based conservation, systematic conservation planning, strategic and opportunistic approaches, and social–ecological ecosystems are some of the important concepts in effectiveness of systematic conservation planning.

Keywords Coupled human–environment systems · Evidence-based conservation · Informed decision-making · Opportunistic conservation approach · Strategic conservation approach · Systematic conservation planning

10.1 Introduction

Conserving biodiversity requires a wide array of resources. Furthermore, understanding the critical constraints mainly ecological, social, and financial constraints would have a critical impact on the success of a proposed conservation project. It is recognized that a significant number of species are on the brink of extinction. However, the budget required to conserve all threatened species is limited globally. This could have a profound influence on the planning methods and conservation strategies of governmental and non-governmental organizations across the globe (Groves et al. 2002). Hence, conservation planning and priorities should consider the aforementioned constraints to achieve the conservation goals. Furthermore, auditing the efficacy of conservation projects is necessary to manage the limited budget allotted for conservation projects. Conservation biology scientists have been emphasizing the need to develop an approach that audits the effectiveness of conservation projects to minimize resource waste. In response to this quest, the concept of "evidence-based conservation" emerged in the twenty-first century.

K. G. Yaynemsa, *Plant Biodiversity Conservation in Ethiopia*,
https://doi.org/10.1007/978-3-031-20225-4_10

Contemporary concepts related to evidence-based conservation and systematic conservation planning are briefly discussed in this chapter.

10.1.1 Evidence-Based Conservation

Understanding the variations in the effectiveness of diverse conservation approaches in spatial and temporal contexts is often not an easy task (Ferraro and Pattanayak 2006). Due to this reason, the need for evidence-based conservation in the twenty-first century is underscored by scholars in the field. The interdisciplinary composition of researchers and conservation experts would have a prominent contribution to evidence-based biodiversity conservation. The experts need to be critical thinkers with a deep understanding of how to make evidence-based decisions (Downey et al. 2021). The traditional, experience and expert-based, decision-making in conservation should be supplemented by evidence-based synthesis to avoid failure of effective management in conservation (Walsh et al. 2014). This is, however, not an easy task. It needs to develop tools that simplify evidence-based decisions and train conservation practitioners and managers (Downey et al. 2021).

The application of informed decision-making is often constrained by the accessibility of the evidence, decision-makers lack of commitment and skill, and conflicting results of the literature (Walsh et al. 2019). Furthermore, institutional culture, decision-making processes, and culture along with practitioners' attitudes and the relationships between scientists and practitioners play a vital role in evidence-based conservation (Walsh et al. 2019). Shreds of evidence in biodiversity conservation could be case reports/studies, experimental, meta-analysis, systematic review, and randomized control trials (Burivalova et al. 2019).

Evidence-based conservation has some core skills. These include ask, access, appraise, apply, and audit (Fig. 10.1) (Young et al. 2014). These core skills require the composition of experts with multidisciplinary expertise. Moreover, the process is not a one-way-go; rather, it is more recurring.

Although the evidence-based decision-making concept is not recent at all globally, scientific evidence is hardly practiced in developing countries such as Ethiopia. The author searched about evidence-based studies in Ethiopia, particularly plant biodiversity conservation, both on the web and organizations that are highly involved in biodiversity conservation. However, such studies are not available. Hence, I recommend the decision-makers review the concept of evidence-based conservation and adopt it provided that they found it important.

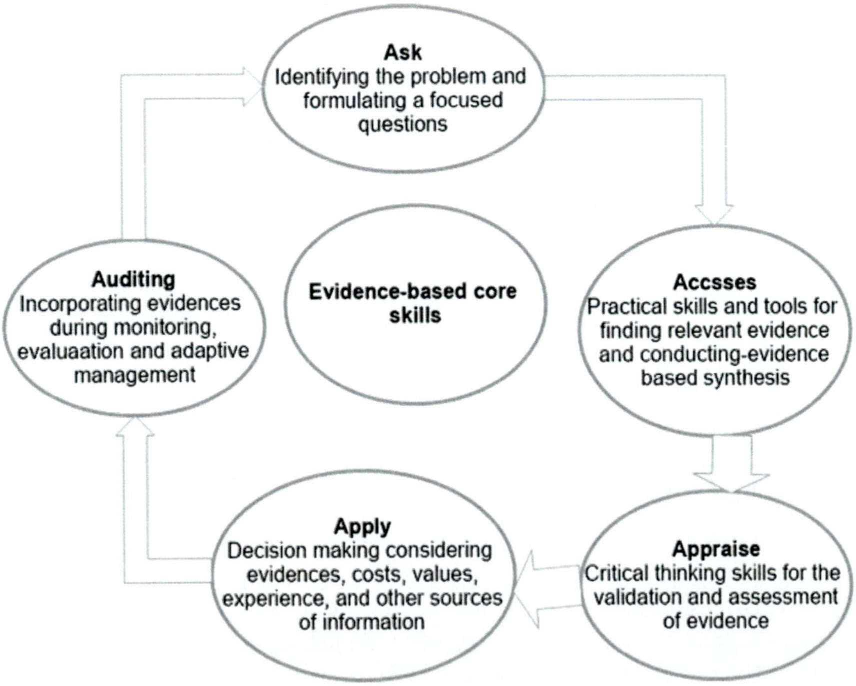

Fig. 10.1 Core skills of evidence-based conservation. Based on (Young et al. 2014; Downey et al. 2021)

10.1.2 Systematic Conservation Planning

Systematic conservation planning (SCP) is a "structured decision analysis approach with a series of stages that concern both the design and implementation of conservation actions to meet specific objectives, usually through the spatial prioritization of the most feasible places for conservation investment" (Villarreal-rosas et al. 2020). It is a scientific approach that is used to find cost-effective sites for nature conservation. The SCP is a concept that was introduced over the last four decades with the notion that although establishing a protected area is the founding principle of conservation it is inadequate for biodiversity conservation (Margules and Pressey 2000; Cowling and Pressey 2003). This planning approach attempts that the conservation of biodiversity to have measurable goals, specific criteria, and flexibility. Ecological connectivity is the crux of this approach. It often suggests cost-effective landscape-level conservation rather than species-level conservation planning. Recently, in addition to nature conservation, incorporating ecosystem services in SCP is becoming a common practice (Villarreal-rosas et al. 2020).

Even though the focus of the earlier concept was on technical tasks, a few years later the socioeconomic component that engages stakeholders' including the lowest administration level through the planning process was incorporated into the

SCP framework (Cowling and Pressey 2003). In the Anthropocene epoch, much of the ecosystems across the globe are transformed. As a result, the natural ecosystems consist very small portion of the terrestrial environment. Thus, conservation experts and practitioners need to brainstorm, implement, and design a representative approach of biodiversity conservation. Furthermore, the multidisciplinary principle of conservation biology should be respected. The stakeholders' involvement starting from project proposal development to implementation and management must be given priority in the SCP framework. In earlier times, SCP proposed a six-stage process (Margules and Pressey 2000) to guide informed conservation decisions. Now, however, there are about 11 stages of SCP (McIntosh et al. 2017; Adams et al. 2019). These are briefly discussed below.

1. *Scope and cost the planning process*

This stage is all about delimiting the planning, i.e., demarcating the boundaries, expertise of the team, the budget available and required for the whole process (Pressey and Bottrill 2009; McIntosh et al. 2017).

2. *Identify and involve stakeholders*

The stakeholders to be involved at all stages of the conservation project—from inception to its implementation must be appropriately identified and consulted (Pressey and Bottrill 2009). Because these stakeholders would have either a direct or indirect impact on the success of the project, most of the time projects that did not consult the target stakeholders are deemed to fail.

3. *Describe the context for conservation areas*

Socioeconomic and political settings that are important for the project are identified and described at this stage. Furthermore, the major threats to biodiversity that can be halted by spatial planning and the broad constraints on, and opportunities for, conservation actions are also described here (Pressey and Bottrill 2009; Adams et al. 2019).

4. *Identify (general) conservation goals*

At this stage, quantitative conservation goals are set. The extent of biodiversity to be conserved (quantity) and the quality of the approaches for instance representativeness and connectivity (quality) should be considered while setting the conservation goals.

5. *Collect data on socioeconomic variables and threats*

This stage is about socioeconomic data collection. The variables might include occupation, extractive uses, costs of conservation, constraints, and opportunities to which planners can respond. This would also involve predictions about the variables that might threaten the planning and implementation processes (Pressey and Bottrill 2009).

6. *Collect data on biodiversity and other natural features*

Biological and ecological data are collected at this phase. For instance, species distribution data, plant functional traits, target species, ecosystem types, ecosystem services, surrogates for biodiversity, and metapopulation dynamics are recorded at this stage (Margules and Pressey 2000; Pressey and Bottrill 2009; McIntosh et al. 2017).

7. *Set (quantifiable) conservation objectives*

SMART objectives, quantitative, qualitative, or both, are established from the already identified goals (stage 4). It is at this stage questions such as "how much," "where," and "when" are described.

8. *Review current achievements of objectives*

Data that is accompanied by field surveys or remote data is used to evaluate the achievement of the conservation objectives set. "The imminence of threat to under-represented features such as species or vegetation types and the threats posed to areas that will be important in securing satisfactory design targets are identified" at this stage (Margules and Pressey 2000). This phase is crucial in implying the gaps encountered in the process.

9. *Select additional conservation areas*

Based on the review result of the previous stage, this stage entails decisions on additional conservation areas that complement the existing ones in achieving objectives (Margules and Pressey 2000; Pressey and Bottrill 2009). This requires the involvement of all stakeholders. Associated factors such as constraints (ecological, financial, and social) and opportunities are recognized. Furthermore, the protected area design is also developed at this stage.

10. *Apply conservation actions to selected areas*

It is the conservation project implementation phase. Several stakeholders are involved at this stage. This needs technical analyses and institutional arrangements to ensure that selected areas are given the most feasible and appropriate conservation management and that areas are prioritized for action when resources are limited (Pressey and Bottrill 2009).

11. *Maintain and monitor conservation areas*

Application of conservation projects after ensuring their feasibility alone is not enough. Although the process is successful, i.e., the objectives set are achieved, continuous monitoring of the conservation area is inevitable to maintain the success. This ensures the long-term goals set and persistence of the values for which they were established (Margules and Pressey 2000; Pressey and Bottrill 2009).

However, SCP is criticized for not considering the complexities associated with the economics of low-income countries (Albers et al. 2016). I rather argue that SCP is one of the important tools that help to manage limited conservation resources provided that it is implemented effectively. This is manifested in several studies

(Margules and Pressey 2000; McIntosh et al. 2017). For example, land acquisition costs, which are the main concern of these scholars, could not be a major constraint in Ethiopia. Because the land policy of Ethiopia states that the government and the people are the owners of the land. Thus, the government is not expected to invest extremely huge money for land acquisition. Nevertheless, the primary criteria for reserve site selection, in developing countries, for instance Ethiopia, are ecological and spatial extent components (Kukkala and Moilanen 2013).

10.1.3 Strategic and Opportunistic Approaches

Conservation approaches could be based on either proactive or reactive. Proactive approaches are often planned before the assessment of the biodiversity status of a particular area. Reactive approaches are usually designed to halt the threats to a threatened species or ecosystem.

The strategic conservation approach is proactive which is designed by considering Geographic Information System (GIS) and remote sensing analysis, landscape ecology concepts, SLOSS theory, and expertise-driven analysis to protect the biodiversity in a target site (Wright 2017). The strategic conservation approach is a decision-making process where decision-makers target landscape-level conservation through the involvement of stakeholders, conservation policy enforcement, and achieving the objectives set by reasonably allocating a budget. Strategic conservation analysis, strategic selection of conservation options, and conservation strategy implementation were proposed as the foundations of strategic conservation approach (Marasinghe 2013) (Fig. 10.2). Even though this approach is costly and time-consuming, its emphasis is on the highest priority areas at a landscape level. However, anthropogenic landscapes which are the result of human disturbances are neglected.

On the other hand, the opportunistic conservation approach is reactive which considers anthropogenic landscapes that have ecological value though not supported by explicit spatial analysis (Wright 2017). The term opportunistic conservation is self-explanatory. It makes use of the available opportunities in a particular site by considering the conservation challenges that characterize this approach (Havlick 2014). This approach needs less time and funds than the strategic approach, in which its success is likely. However, it does not focus on high-priority areas.

10.1.4 Drawbacks and Improving Effectiveness of Systematic Conservation Planning

Applications of SCP that target one ecosystem service or beneficiary group will lead to trade-offs across space and time for other non-targeted services, beneficiaries, and biodiversity (Villarreal-rosas et al. 2020). Therefore, SCP should be holistic in design.

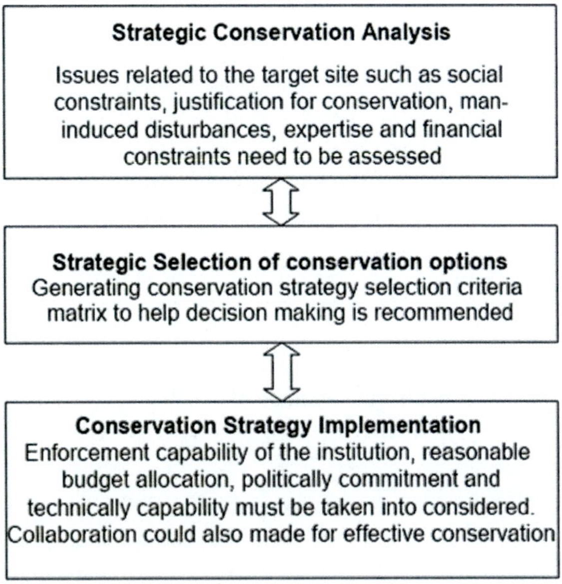

Fig. 10.2 Foundations of strategic conservation planning. Based on Marasinghe (2013)

The following could be some of the constraints that may hamper the implementation of SCP in the Global South in general and Ethiopia in particular.

Institutional setting: institutions that enforce SCP are not appropriately established in Ethiopia at all administration levels, i.e., from Kebelle (the smallest administration level) to the national level. Furthermore, most of the institutions involved in plant biodiversity conservation rely on the traditional conservation discourse. The Return-On-Investment (ROI) or the economic perspective of protected areas is often neglected.

Socioeconomic setting: Ethiopia, with a nearly 120 million population, is the second populous country in Africa following Nigeria. It is predicted to increase largely in the coming decades. This ever-increasing population demands resources including land. Even though the land is owned both by the people and by the government in Ethiopia, usually the willingness of the people to give land for protected area conservation could be one of the most challenging issues. It is because the Ethiopian people are dominantly dependent on the land, often for agriculture. However, it is not only the people but the land is the main capital for the government too. Hence, the social acceptance of conservation projects in low-income countries such as Ethiopia could be a constraint. The involvement of the local people in the SCP process could reasonably minimize this problem.

Lack of appropriate expertise: The composition of experts with multidisciplinary knowledge and technical skills on SCP is extremely limited in the institutions established to enforce conservation approaches. Thus, designing a team composed of experts with some fundamental knowledge on SCP would have a crucial contribution to SCP.

Public awareness: Conservation projects' cost is mainly covered by the government. That means it is the taxpayers' money that is invested in protected area conservation. However, the public rarely requests clarification about the effectiveness of conservation projects and often does not ask for alternative and effective approaches that reduce conservation costs. Most of the time, nobody could be held accountable for failed conservation projects. In order to use the available limited budget for conservation and improve conservation success, the public should request effective and contemporary conservation approaches such as SCP- or evidence-based conservation.

10.1.5 Social–Ecological Systems (SES) or Coupled Human–Environment Systems (CHES)

Traditional conservation theories and practices prioritized the biodiversity and environmental perspective. Consequently, the conservation reserves were mostly established in a relatively undisturbed ecosystem and are supposed to remain intact for a longer time. The social and economic perspectives received little attention. However, conservation projects that ignore the social aspect in the process would never be successful. This is because man-induced disturbances in the Anthropocene threaten the biodiversity in conservation reserves and the status of the ecological systems affects human beings at the local and regional level. As a result, conservation reserves are complex systems that couple human and environmental aspects, which interact with biological, socioeconomic, cultural, and political perspectives at different spatial scales (Cumming and Allen 2017). Hence, the major threats to biodiversity in the conservation reserves (Chap. 5) and the level of vulnerability to the threats must be considered while designing social–ecological resilient (Cumming et al. 2015; Cumming and Allen 2017). Janzen (1983) precisely stated this phenomenon as protected areas could not be considered as purely "ecological islands."

As a response to this demand, the social–ecological systems (SESs) concept is recently recognized as interdisciplinary research. Social ecosystems are complex systems that at least consist of interacting attributes such as socioeconomic, ecological, and political which are useful for conservation reserve management through the involvement of managers, policy- and decision-makers, institutions, and other stakeholders (Cumming et al. 2015). This concept helps to understand the threats and vulnerability of conservation reserves and draw management options effectively. Scholars have developed several SESs frameworks.

Ostrom developed an SES framework (Fig. 10.3) that comprised four core components, namely resource systems, resource units, actors, and governance systems (Ostrom 2007, 2009). Its application in ecology is, however, criticized because it lacks clear definitions concerning resource units and resource systems; and is poorly equipped to analyze biophysical processes and diagnose ecological contributions to social–ecological outcomes (Cumming et al. 2015). Furthermore, in order to promote sustainable development through sustainable utilization, scholars developed some social–ecological system frameworks that consist: identifying hotspots of ecological functions and fragility, applying theory into practice, and integrating local communities in developing management policy of conservation reserves (Zhang et al. 2020). SES approach demands cooperative activities, flexibility, humbleness, and trust of stakeholders in biodiversity conservation in conservation reserves (Steger et al. 2021).

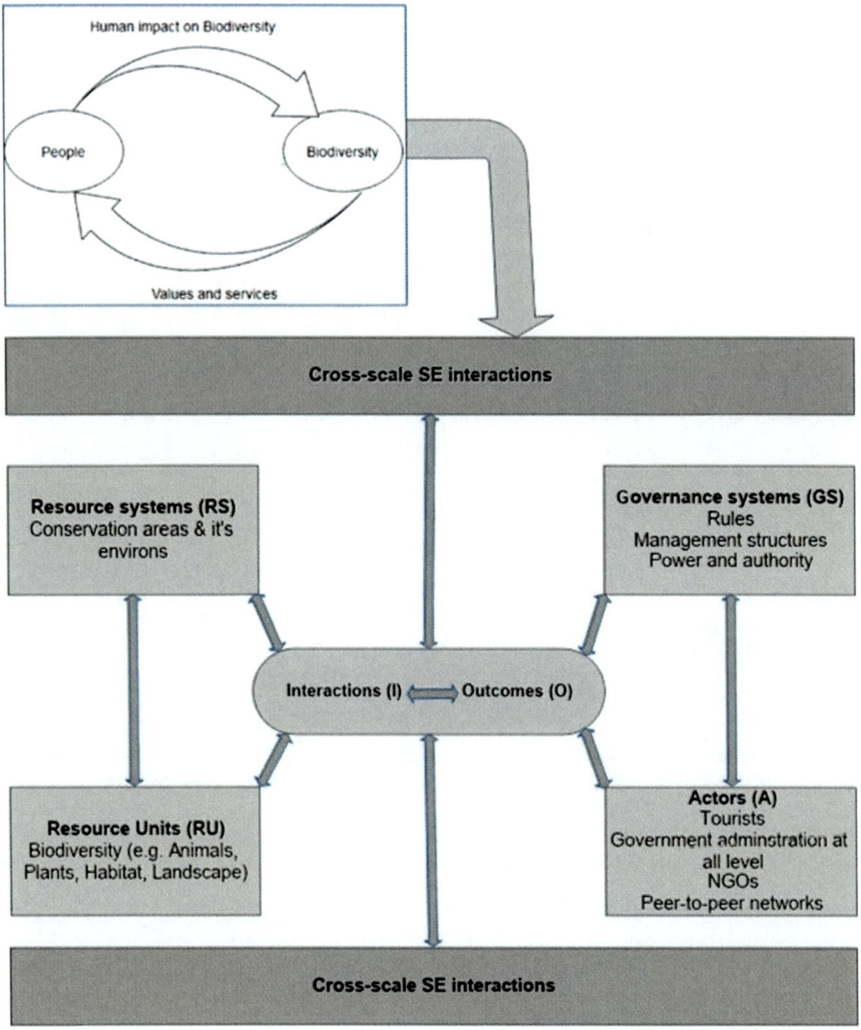

Fig. 10.3 Ostrom's social–ecological systems framework (Ostrom 2007, 2009). The social–ecological systems (RS, RU, GS, and A) interact to produce outcomes. The interactions are by no means unidirectional. RU is part of the RS and is inputs to the focal action situations (I & O). The RS set conditions for the I & O. The GS on the other hand defines and sets rules for actors to participate in the I & O

Recently, three themes are emerging in SESs approaches to understanding conservation reserves (Cumming and Allen 2017). These are—the resilience and sustainability of conservation reserves, the relevance of spatial context and scale for conservation reserves, and the relationships of people and nature. The resilience and sustainability of conservation reserves are related to the ability of an ecosystem to withstand anthropogenic and natural disturbances and remain in a slightly changing state, which implies the dynamics of change as well as the effectiveness of the protected area. Spatial context and scale are about the geographic boundary of the conservation reserves and how the ecological connectivity or networks are treated from the conservation reserve perspectives. The people and nature relationship is an age-old concept that is often associated with ecosystem services delivered and anthropogenic disturbances on biodiversity.

However, SESs research should not be considered as problem-solving. Rather, its broader goal is to generate interdisciplinary knowledge that can be used by several stakeholders policy and decision-makers, NGOs, and agencies to promote long-term scientific knowledge and evidence into policy processes (Turner et al. 2016). Thus, researchers might incorporate as many components and sub-components as possible to extract the most important evidence for biodiversity conservation by promoting sustainable development. If researchers fail to address the main barriers in the SESs, the intended conservation target would not be fully achieved. A team of multidisciplinary researchers would crucially contribute to SESs research.

References

Adams VM, Mills M, Weeks R et al (2019) Implementation strategies for systematic conservation planning. Ambio 48:139–152. https://doi.org/10.1007/s13280-018-1067-2

Albers HJ, Maloney M, Robinson EJZ (2016) Economics in systematic conservation planning for lower-income countries : a literature review and assessment. Int Rev Environ Resour Econ 10:145–182. https://doi.org/10.1561/101.00000085

Burivalova Z, Miteva D, Salafsky N et al (2019) Evidence types and trends in tropical forest conservation literature. Trends Ecol Evol 34:669–679. https://doi.org/10.1016/j.tree.2019.03.002

Cowling RM, Pressey RL (2003) Introduction to systematic conservation planning in the Cape Floristic Region. Biol Conserv 112:1–13. https://doi.org/10.1016/S0006-3207(02)00418-4

Cumming GS, Allen CR (2017) Protected areas as social-ecological systems: perspectives from resilience and social-ecological systems theory. Ecol Appl 27:1709–1717. https://doi.org/10.1002/eap.1584

Cumming GS, Allen CR, Ban NC et al (2015) Understanding protected area resilience: a multi-scale, socialecological approach. Ecol Appl 25:299–319

Downey H, Cooke SJ, Walsh JC et al (2021) Training future generations to deliver evidence-based conservation and ecosystem management. Ecol Solut Evid 2:e12032. https://doi.org/10.1002/2688-8319.12032

Ferraro PJ, Pattanayak SK (2006) Money for nothing? A call for empirical evaluation of biodiversity conservation investments. PLoS Biol 4:0482–0488

Groves CR, Jensen DB, Valutis LL et al (2002) Planning for biodiversity conservation : putting conservation science into practice. Bioscience 52:499–512

Havlick DG (2014) Opportunistic conservation at former military sites in the United States. Prog Phys Geogr 38:271–285. https://doi.org/10.1177/0309133314522281

Janzen DH (1983) No park is an island, increase in interference from outside as park size decreases. Oikos 41:402–410

Kukkala AS, Moilanen A (2013) Core concepts of spatial prioritisation in systematic conservation planning. Biol Rev 88:443–464. https://doi.org/10.1111/brv.12008

Marasinghe MSLRP (2013) Strategic conservation: a review of protected area management. J Environ Prof Sri Lanka 2:66–76

Margules CR, Pressey RL (2000) Systematic conservation planning. Nature 405:243–253

McIntosh EJ, Pressey RL, Lloyd S, et al (2017) The impact of systematic conservation planning. Annu Rev Environ Resour 42:677–97. https://doi.org/10.1146/annurev-environ-102016-060902

Ostrom E (2007) A diagnostic approach for going beyond panaceas. Proc Natl Acad Sci U S A 104:15181–15187

Ostrom E (2009) A general framework for analyzing sustainability of social-ecological systems. Science (80-)352:419–422

Pressey RL, Bottrill MC (2009) Approaches to landscape- and seascape-scale conservation planning: convergence, contrasts and challenges. Oryx 43:464–475. https://doi.org/10.1017/S0030605309990500

Steger C, Klein JA, Reid RS et al (2021) Science with society: evidence-based guidance for best practices in environmental transdisciplinary work. Glob Environ Chang 68:102240. https://doi.org/10.1016/j.gloenvcha.2021.102240

Turner BL II, Esler KJ, Bridgewater P et al (2016) Socio-environmental systems (SES) research: what have we learned and how can we use this information in future research programs. Curr Opin Environ Sustain 19:160–168. https://doi.org/10.1016/j.cosust.2016.04.001

Villarreal-rosas J, Sonter LJ, Runting RK, et al (2020) Advancing systematic conservation planning for ecosystem services. Trends Ecol Evol 35:1129–1139. https://doi.org/10.1016/j.tree.2020.08.016

Walsh JC, Dicks LV, Raymond CM, Sutherland WJ (2019) A typology of barriers and enablers of scientific evidence use in conservation practice. J Environ Manage 250:109481. https://doi.org/10.1016/j.jenvman.2019.109481

Walsh JC, Dicks LV, Sutherland WJ (2014) The effect of scientific evidence on conservation practitioners' management decisions. Conserv Biol 29:88–98. https://doi.org/10.1111/cobi.12370

Wright JB (2017) Comparing strategic and opportunistic approaches to land conservation: a case study of the new Mexico land conservancy. Pap Appl Geogr 3:101–109. https://doi.org/10.1080/23754931.2016.1148068

Young T, Rohwer A, Volmink J, Clarke M (2014) What are the effects of teaching evidence-based health care (EBHC)? Overview of systematic reviews. Plos One 9:e86706. https://doi.org/10.1371/journal.pone.0086706

Zhang K, Gao J, Zou C et al (2020) Expansion of protected area networks integrating ecosystem service and social-ecological coordination. Glob Ecol Conserv 24:e01298. https://doi.org/10.1016/j.gecco.2020.e01298

Chapter 11
Trends in Plant Ecology Research in Ethiopia

Abstract The progress of plant ecology research in Ethiopia is not known to date. Consequently, it is challenging to judge the impact of existing research on policy development and conservation actions. The objective of this paper was to systematically analyze the trend of plant ecological research in Ethiopia over the past 50 years. The inclusion and exclusion of the articles for analysis were performed using Reporting Standards for Systematic Evidence Syntheses (ROSES) flow diagram developed for systematic review/meta-analysis. The number of articles published, authors, and collaboration has increased dramatically since the 1960s. Most of the research (52.6%) focused on the Dry evergreen Afromontane Forest and grassland complex (DAF) and Moist evergreen Afromontane Forest (MAF) vegetation types. Of the remaining vegetation types, the woodlands (14.3%), i.e., *Acacia-Commiphora* woodland and bushland proper (ACW), and *Combretum-Terminalia* woodland and wooded grassland (CTW), desert and semi-desert scrubland (DSS) (2.3%), and the Afroalpine (AA) and Ericaceous Belt (EB) (1.5%) received comparatively little attention. Classical plant ecology themes, descriptive plant community studies, were dominant over the last five decades in contrast to contemporary themes globally. Other plant ecological studies such as reproductive and dispersal ecology of invasive plant species, and pollination ecology seem to be largely neglected topics. Furthermore, the recommendations forwarded by most of the articles reviewed (38.1%, $n = 51$) were not result-based. As a future direction, the Ethiopian government should develop a project database for both completed and ongoing projects.

Keywords DAF · MAF · Research syntheses · ROSES · Systematic review · Vegetation ecology

11.1 Introduction

Plant ecology as a standalone discipline of botany has a long history. It is highly linked to the works of Alexander von Humboldt in the early nineteenth century

Reprinted with slight modification from BOTHALIA—African Biodiversity and Conservation, 52(1) (2022). DOI: https://doi.org/10.38201/btha.abc.v52.i1.12

K. G. Yaynemsa, *Plant Biodiversity Conservation in Ethiopia*,
https://doi.org/10.1007/978-3-031-20225-4_11

(Hagen 2010). Subsequently, some branches of plant ecology emerged, such as synecology and autecology, which place emphasis on community ecology and individual species, respectively. From the early nineteenth century onwards, plant ecologists have studied stands of vegetation, which they considered samples of a plant community (Mueller-Dombois and Ellenberg 1974). Currently, however, traditional ecological terms (e.g., synecology and autecology) are replaced by specialties such as population ecology, community ecology, ecosystem ecology, ecological modeling, global change biology, and remote sensing (Hagen 2010; Asselin and Gagnon 2015; Grace 2019).

Even though plant ecology research, which focused on plant communities, had been the center of ecological research during the first half of the twentieth century (Hagen 2010), plant ecology as a discipline has changed as the scope of the research themes has grown over-time. However, there are arguments about the evolution of plant ecology research. For example, Peters (1991), O'Connor (2000) criticized and argued that ecology as science has not grown up rather the science moves forward slowly. On their critical response to Peters' criticism and arguments, Grace (2019), Nobis and Wohlgemuth (2004) countered that ecological research is growing both in scope and citation impact as well. It is believed that advances in population genetics and the evolutionary theory shifted the theme of plant ecological research into a broader scope, which includes population ecology that combined mathematical modeling and experimentation to investigate population growth, dispersal, and competition from an explicitly Darwinian perspective (Harper 1967, 1977; Hagen 2010; Asselin and Gagnon 2015; Grace 2019). For instance, McCallen et al. (2019) identified nearly 50 research topics in ecology over the past four decades.

Like the research topics, the methods employed in plant ecology research also vary depending on the objectives set. These might vary in terms of spatial and temporal scale, organizational levels such as species, population, community, and ecosystem. Furthermore, the approaches could be either classical or advanced (Henderson 2012). Although the use of some classical approaches could be retained, several plant ecology research methods are being evolved. Consequently, research in plant ecology is supported by several software systems and becoming more reputable. This could have a substantial contribution to vegetation management and biodiversity conservation.

Ethiopia is a country with a very complex topography with elevation varying from about 125 m below sea level to about 4533 m above sea level (m a.s.l.) (Gebrehiwot et al. 2020). Two of the 36 biodiversity hotspots: the Eastern Afromontane and the Horn of Africa are found in Ethiopia (Mittermeier et al. 2004; Hoffman et al. 2016; Gebrehiwot et al. 2020). Thus, Ethiopia is regarded as the major center of diversity and endemism for several plant species though there are extensive anthropogenic disturbances. Taking the complex topography and diverse land-use types into consideration, the country has the potential to offer a myriad of research opportunities in plant ecology—across a range of themes. Some vegetation surveys were performed across northeastern Africa (Ethiopia, Eritrea, Djibouti, and Somalia) in the late 1950s (Pichi-Sermolli 1957).

According to the information extracted from different sources, however, empirical plant ecology research in Ethiopia started in the late 1960s. The objective of, presumably, the first empirical plant ecology research in Ethiopia was to test community ecology hypotheses (Beals 1969). Since Beals' publication, significant numbers of plant ecological research have been conducted. However, there is no empirical data on the trend of plant ecology researches in Ethiopia. As a result, the progress of plant ecology research in the country is yet unknown. It is, thus, challenging to understand the progress and impact of these researches on policy development and conservation actions. Therefore, a thorough bibliographic analysis of plant ecology research in Ethiopia is inevitable to understand if the discipline is growing up as a science in the region. This could have a crucial role in showing what has been done so far and what should be in this field.

The objective of this paper is to systematically analyze the trend of plant ecological research in Ethiopia with the aim of providing responses to the questions: (i) what are the most researched vegetation types (i.e., type of natural ecosystem, such as forests) and land-use types (i.e., the purpose for which the land is used such as farmland), (ii) what are the most researched domains of plant ecology, (iii) is plant ecology research influencing national plant biodiversity conservation policy, and (iv) are there ecological research funding from the government and international sources?

11.1.1 Authorship and Collaboration

Vegetation ecology research in Ethiopia revealed a linear increase in publication ($P < 0.001$, R square = 0.63) (Fig. 11.1a). However, there have been interruptions between 1960 and 1990s. There has been a dramatic increase after the 1990s ($P < 0.001$, R square = 0.60). The articles published in 1969 and 2018 were 2 and 27, respectively. Similarly, the number of authors also showed an increasing trend (Fig. 11.1b). While the lowest number of authors was recorded as one in different years, the highest number of authors (24) was recorded in 2018. This revealed that collaborative research showed an increasing trend over the study period.

The author/s affiliation/collaboration involved showed local authors are dominant, 83 (62%). On the other hand, international collaborators and foreign authors only comprised 48 (35.8%) and 3 (2.2%), respectively. International collaborations started in the late 1990s and grew steadily. On the contrary, the research conducted by foreign nationals only is negligible. The only articles published solely by the foreign nationals were in 1969 and 2017.

The low number of plant ecological studies in the late 1960s is associated with the lack of trained manpower and political instabilities in the country. Exclusive to Haramaya and Gondar, which were colleges of agriculture and health science, respectively; there has been only one university with a natural science program, the so-called Haile Selassie I University, current Addis Ababa University since the early 1970s. Furthermore, most of the experts at the university were foreigners with

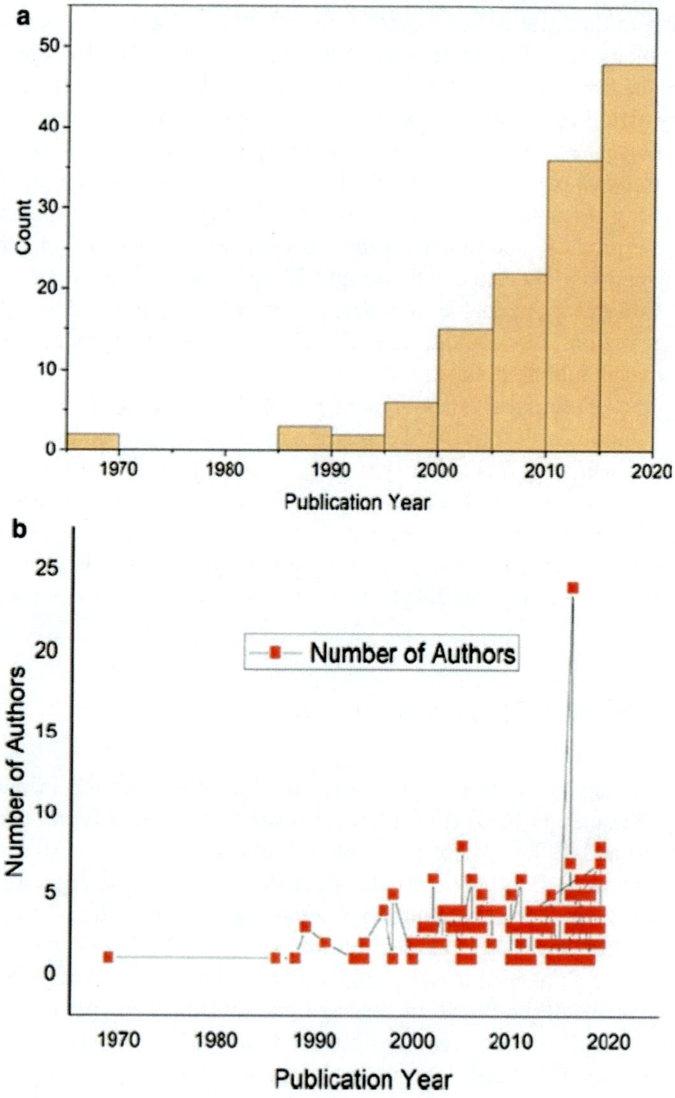

Fig. 11.1 Trend of plant ecology research in Ethiopia. **a** Number of published articles and **b** trend of plant ecological research collaboration

limited knowledge about the vegetation of Ethiopia. Hence, the small number of plant ecology research articles is expected since the early 1970s.

However, from the early 1980s, thanks to the flora project funded by Swedish International Development Cooperation Agency (SIDA), experts were trained on several disciplines such as plant ecology, plant systematics, and ethnobotany. A few of them are still staff members of Addis Ababa University. Furthermore, with the number of universities in the country increasing from one in the 1960s to nearly 50 today. This has tremendously increased the number of trained manpower. Concurrently, the number of publications rose steeply in the last five decades. In addition to the increased number of researchers over the last five decades, the increase in the number of subjects due to new approaches and the development of new technologies from different disciplines have influenced the increase in plant ecology research publications (Kim et al. 2018).

Globally, the decline of single-authored articles has been confirmed in various disciplines (Barlow et al. 2018). Similarly, in the present study, single-authored articles showed a declining trend. This is mainly due to the fact that the research projects are either multidisciplinary or students and staff resulting with multiple authorship articles produce better quality research. Furthermore, most of the research are theses/dissertation outputs which include theses/dissertation supervisors and students contributing to multi-authored articles.

11.1.2 Plant Ecological Research Components

Results revealed that most of the articles' objectives were descriptive (Table 11.1). A few articles dealt with some advanced objectives. About 42 (31%) of the articles' objectives were on the floristic survey, community structure analysis, and assessing the regeneration status of a forest based on a seedling, sapling, and mature tree count. Whereas, the Theory-Approach development OR Life form/Functional Traits OR Climate change/Sustainability theme covered only 5 (3.7%) of the articles. Even though few articles are available about invasive plant species distribution, and economic impact, studies on reproductive and dispersal ecology of invasive species were almost non-extant. Furthermore, pollination ecology seems the neglected topic in the Ethiopian academia. Exclusive to the very few articles on crop pollination such as Coffee (*Coffea arabica*) by honeybee (*Apis mellifera*), pollination ecology on indigenous and wild plants is badly missing.

Table 11.1 Thematic research topics in the articles

Objectives	Frequency	Percent (%)
Floristic, structure, regeneration (count)	42	31.3
Vegetation-environmental-disturbance relationships OR carbon estimation	35	26.1
Floristic, structure, SSB	28	20.9
Floristic (woody/herbaceous)	16	11.9
Vegetation-environmental-disturbance relationships, GIS and RS	8	6.0
Theory-approach development or life form/functional traits or climate change/sustainability	5	3.7
Total	134	100.0

Out of the 134 articles reviewed, only two articles were experimental while the remaining are descriptive. One of these articles investigated the tree regeneration potential of only four species namely *Juniperus procera*, *Ekebergia capensis*, *Prunus africana,* and *Olea europaea* subsp. *cuspidata* under three conditions, i.e., along the interior and edge forest gradients, canopy cover, and grazing intensity (Wassie et al. 2009). The other article determined the floristic composition and soil seed bank richness using manure and livestock grazing as treatments (Woldu and Saleem 2000). The objectives of these two articles were in a similar domain.

In the twenty-first century, research is supported by many technology products such as equipment for experiments, software programs for statistical analysis, and other applications such as GIS and remote sensing which support field works. Consequently, the number of plant ecology articles published and the research theme is dramatically increasing globally (McCallen et al. 2019). In the present systematic review, classical plant ecology themes were dominant over the last five decades in contrast to contemporary themes such as microscale, macroscale, and anthropogenic themes globally (McCallen et al. 2019), which employs sophisticated techniques of analysis, including multivariate statistics and mathematical modeling (Kim et al. 2018). McCallen et al. (2019) identified nearly 50 thematic areas in ecological research. However, less than five thematic areas were covered in the articles reviewed in this study. Descriptive studies on plant community ecology revolving on floristic composition and community structure dominated the studies. However, the few studies on invasive species, fire ecology and species distribution modeling in the twenty-first century reveals that plant ecological research is in a way of transition into a new research theme. The dominance of community ecology research was reported by Réale et al. (2020) in their review based on co-citation analysis although it was in the 1970s. This is, however, in contrast to a report by Carmel et al. (2013) in which single species research found to be dominant.

Ethiopia's natural environment is extremely degraded (Wassie 2020). As a result, most of the environment in the country is a fertile ground to invasive alien, particularly, plant species. Numerous plant species have been recognized as invasive in Ethiopia. Although most of them are exotic, there are few native plants which turned into invasive. Some of these invasive species appear to have become established in the desert and semi-desert scrubland at elevation lower than 1000 m a.s.l. (e.g., *Parthenium hysterophorus* and *Prosopis juliflora*) and Afromontane forests at elevation above 2500 m a.s.l. (*personal observation*), suggesting that protected areas in the lowlands and highlands are prone to invasive species. However, few studies investigated the distribution and impact of invasive species (Shiferaw, 2019; Shiferaw et al. 2019a, 2019b). Even though the study on invasive species distribution and cover is extremely important for its management, understanding the reproductive and dispersal ecology of the species also play a crucial role either for management or eradication of the undesirable invasive species. Their impact is profound in the rift valley. However, studies on this topic, as well as in this ecosystem are extremely limited. The persistent environmental degradation, increasing trend of invasive species, and climate change are aggravating the impact of invasive species. Hence, research on this field is critically in need.

Pollination, often-mutualistic interaction, is at the core of ecological networks, which plays a vital role in maintaining community stability and ecosystem function. Nevertheless, these interactions are threatened due to natural and anthropogenic disturbances leading to a global decline of pollinators. Plant–pollinator networks can potentially modify the population dynamics and the occupied range of a plant species (Pellissier et al. 2012). However, the contribution of pollination networks as driver of plant distribution and assemblage of plant communities has received little attention. Thus, there are a strong need to characterize plant–pollinator interactions at large spatial scales and especially with respect to dynamic communities, whose compositions and patterns of relative species abundance vary in time and space (Pellissier et al. 2012).

In the Anthropocene, global warming is causing changes in species fundamental and realized environmental niches. As a result, plants are shifting their ranges and phenology. Consequently, this causes a plant–pollinator mismatches which leads to plant and pollinators diversity decline. The most species at risk could be plants that required specialized pollinators while the generalized ones are least affected. This is due to specialized plants will be more prone to pollen limitation because they are less likely to interact with any available pollinator (Knight et al. 2005). The impact of low abundance and diversity of pollinators do not only influence the abundance of individual plant species but do extend to shaping plant communities.

Apart from the scant pollination studies on crops (Samnegård et al. 2014, 2016; Geeraert et al. 2020), particularly *Coffea arabica*, research on plant–pollinator networks in natural plant communities is absent in Ethiopia. This is identified as an important gap. Hence, future plant ecology research in Ethiopia should consider pollination ecology.

The limited number of articles from experimental studies and advanced themes could be due to the lack of resources for experimental activities and limited expertise in these fields. Unlike the present study, Carmel et al. (2013), Asselin and Gagnon (2015) revealed both observation and experimental studies shared almost equal contributions in their systematic review.

Although the Swedish organization (SIDA) contributed to training experts from different fields, the themes were bound to basic sciences. Hence, these experts have worked on descriptive research, and the encourage the students they supervised them to work their theses/dissertation on similar topics. Once, the graduates are distributed to different universities, they follow in the footsteps of their supervisors. As a result, redundancy of research thematic occurs throughout the country.

11.1.3 Plant Ecological Research on Vegetation and Land-Use Types

Most of the plant ecological research articles in Ethiopia focused on the DAF and MAF vegetation types (Figs. 11.2 and 11.3). Research on these vegetation types

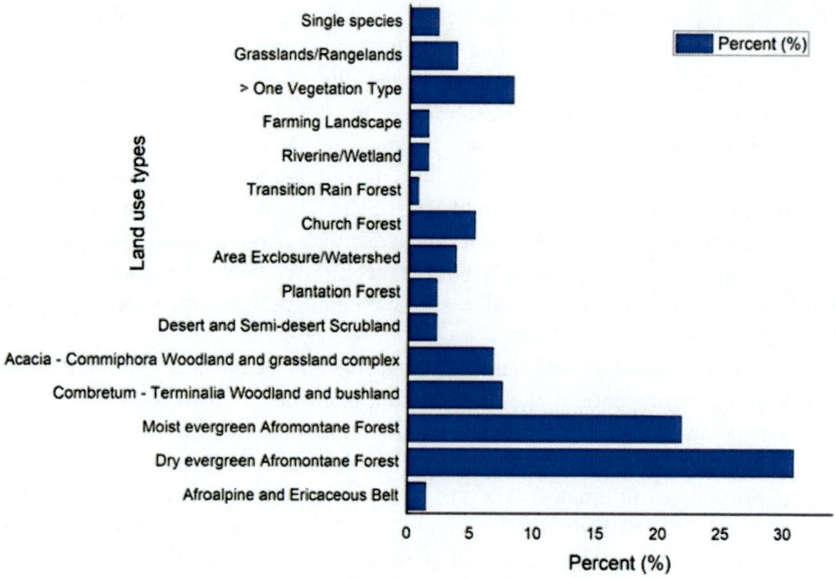

Fig. 11.2 Distribution of the studies in vegetation types and land uses

comprised about 52.6%. However, the woodlands (14.3%), i.e., *Acacia-Commiphora* woodland and bushland proper (ACB), and *Combretum-Terminalia* woodland and wooded grassland (CTW), desert and semi-desert scrubland (DSS) (2.3%), and the threatened Afroalpine (AA) and Ericaceous Belt (EB) (1.5%) received little attention. The Transitional Rain Forest (TRF) vegetation type was represented by only one article. Furthermore, nearly 8.3% of the studies covered more than one vegetation type. The Church Forests, grasslands/rangelands, area exclosures comprised 5.3%, 3.8%, and 3.8%, respectively. Apart from the natural vegetation types, other land uses have also been an area plant ecological research. Plant ecology research on farmland landscape and plantation forests comprised 1.5 and 2.3%, respectively. About 2.3% of the studies were focused on single species. Nevertheless, the absence of appropriate geographical coordinates makes tracing some of the study sites extremely challenging.

11.1.4 Plant Community Types

About 43% of the studies reported plant community types analysis. The number of communities varied from 2 to 9 with a mean of 5 communities. However, some articles did not follow the standard community (*Juniperus – Olea* community) naming

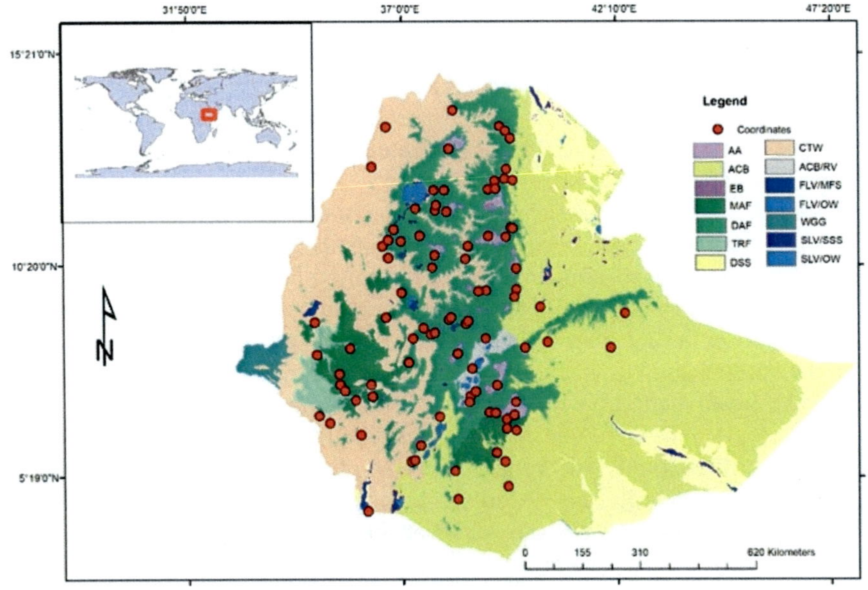

Fig. 11.3 Map of the studies with formal coordinates (red dots) and the corresponding vegetation types (AA = Afroalpine Belt, ACB = *Acacia-Commiphora* woodland and bushland proper, EB = Ericaceous Belt, MAF = Moist evergreen Afromontane Forest, DAF = Dry evergreen Afromontane Forest, TRF = Transitional Rain Forest, DSS = Desert and semi-desert scrubland, CTW = *Combretum-Terminalia* woodland and wooded grassland, ACB/RV = *Acacia* wooded grassland of the Rift Valley, WGG = Wooded grassland of the Western Gambela region, FLV/MFS = Freshwater marshes and swamps, floodplains and lake shore vegetation, FLV/OW = Freshwater lakes—open water vegetation, SLV/SSS = Salt pans, saline/brackish and intermittent wetlands and salt-lake shore vegetation, and SLV/OW = Salt lakes—open water vegetation) (Friis et al. 2010; van Breugel et al. 2015)

procedure while others showed some deviation in the characteristic species of vegetation types. For example, *Olinia rochetiana* (Oliniaceae) is described as a characteristic species of the DAF (Friis et al. 2010). However, this species was reported as a characteristic species of MAF. Furthermore, although *Erica arborea* is a characteristic species of Ericaceous Belt, about four articles reported it as a community of the DAF. Furthermore, a shrub/tree and herb (e.g., *Albizia schimperiana—Hypoestes forskaolii*, *Hyparrhenia filipendula—Combretum molle*) were frequently used to name a plant community. Others also named plant community after a weed such as *Achyrantes aspera*. Overlap of plant community types between vegetation types was also reported in different articles. For example, *Arundinaria alpina* and *Maesa lanceolata—Brucea antidysenterica* communities were reported both from DAF and MAF.

Performing ecological research is not an easy task. Taking ecological data from the field requires physical fitness, withstanding harsh conditions, and other field phenomena. Most of the plant ecological research was in either the DAF or MAF vegetation types. This could be due to two main reasons. First, the highest priorities were given to forests and most of the forest cover in Ethiopia is found in these vegetation types. Second, the DAF and MAF are found in a relatively suitable climate where most of the population is currently residing. Furthermore, they are easily accessible. As a result, these are the most threatened vegetation types (Woldu 1999) which needs empirical research.

However, the other important vegetation types attracted little attention. Particularly, the Afroalpine and Ericaceous belt, which are the main sources of freshwater for the downstream population unexpectedly, received little attention. These could be due to either the unwelcoming environmental conditions such as extreme cold and inaccessibility or the little priority given to these vegetation types. Similarly, articles from the vegetation types from drylands/lowlands contributed little. Furthermore, it could be due to the fact many believe that "*Everything is in and from the forest*" syndrome. Many think that if forests are restored, biodiversity/natural resource would be restored, which is not always true. According to a study that analyzed about 5000 tweeter posts/tweets, this disparity is reflected at global level (Silveira et al. 2021). The ecosystem services and values of forests and open ecosystems (Bond 2019; Silveira et al. 2021) may not always be similar. Hence, researchers and policy- and decision-makers should always recognize open ecosystems such as grasslands, shrublands, and woodlands.

11.1.5 Methods Employed by the Studies

The selection of data collection and analysis methods might be based on the availability of time, fund, expertise, and objectives among others. In the present review 86 (64.2%) of the studies employed systematic sampling while a combination of sampling methods and plot-based data collection supported by GIS only accounted 6 (4.4%) studies. The analysis methods also revealed above 50% of the articles were descriptive (Table 11.2).

11.1.6 Recommendations Forwarded by the Studies and Funding Sources Reported

Most of the studies' recommendations were not based on the results of the research (38.1%, $n = 51$). For example, some floristic composition studies recommended the establishment of "Natural reserve" or "Biosphere Reserve." Although floristic study is part of biosphere reserve, proposing "Natural reserve" or "Biosphere Reserve" based on solely floristic list is far beyond the minimum requirement. Only 21.6% ($n = 29$) of the results revealed that the recommendations were based on the results. On the other hand, 16.4% ($n = 22$) of the recommendations showed the gaps that were

Table 11.2 Descriptive statistics of the sampling and analysis method of the studies reviewed

Sampling method			Analysis method		
Description	Frequency	Percent (%)	Description	Frequency	Percent (%)
Systematic	86	64.2	Descriptive and community classification	34	25.4
Not mentioned	14	10.4	Descriptive	31	23.1
Preferential	14	10.4	ANOVA and model	25	18.7
Stratified	6	4.5	Descriptive, community classification, and ordination	24	17.9
Random	5	3.7	Ordination and GIS/RS	15	11.2
Combined	5	3.7	Descriptive, community classification, ordination plus socio-economic	5	3.7
Plot and GIS	1	0.7	Total	134	100
Total	134	100			

not covered in their studies. On the other hand, several studies (23.9%, $n = 32$) did not imply any recommendation.

Plant ecological research often-forward recommendations based on the empirical data and conclusions made. This serves as an input for natural resource managers and decision-policy makers in conservation practices. The stakeholders decide based on the concrete data and recommendations made by the researchers. Thus, its impact is often manifested in conservation practices. Furthermore, showing the thematic research gap would also have profound benefits. In the present study, most of the recommendations were not made based on the research results despite their relevance to decision-makers. This in some way shows either missed opportunity to help the stakeholders or obscure research objective. However, the authors of this paper are cognizant forwarding recommendation may not be always necessary.

Furthermore, monitoring of the completed projects and incorporating the research recommendations into policy- and decision-making is poorly noticed. For example, Teketay and Bekele (1995), Hundera and Deboch (2008), Hundera et al. (2007) recommended Wof-Washa, Gurra Farda, and Dodola forest as a nature reserve. However, they are not designated as a protected area yet (UNEP-WCMC 2017). Furthermore, forest and woodland cover loss are reported even in protected areas (Nune et al. 2016; Arafaine and Asefa 2019), revealing loose policy- and decision-making. These could be due to three reasons. These are (i) the government is responsible for integrating the research outputs into policies, (ii) complexity of socio-ecological systems (Miteva 2012), lack of funding for conservation operations (Ferraro and Pattanayak 2006), and (iii) lack of trustworthy scientific studies.

The funding source for the articles reviewed was dominantly from international funders 43.3% ($n = 58$) and government 35.1% ($n = 47$). The remaining articles' funding sources were either collaborative (8.2%, $n = 11$) or did not mention (13.4%, $n = 18$) their funder. It is believed that international funding enables researchers to conduct a study with breadth. Although the international funding sources have made a significant contribution, the thematic of most of the articles was descriptive as shown in the previous sections.

Whether the funding source is governmental or international, the funders might have their own interest. Oftentimes, funders are interested in making conservation decisions based on scientific evidence (Burivalova et al. 2019), to maximize the beneficial outcomes of conservation, given the limited resources available (Game 2018). It is interesting that international funding dominated in the studies reviewed. However, the research components investigated by using international funding are more or less similar to the investigations conducted by government funding.

11.2 Is Plant Ecology Research Growing Up in Ethiopia?

There are debates on the progress of ecological research. Some scholars argue ecological research is showing slow progress (Peters 1991; O'Connor 2000) and others claim it is positively progressing by providing some shifts in research themes (Nobis and Wohlgemuth 2004; Grace 2019). Bringing Grace's question, "is ecology growing up?" (Grace, 2019) to the Ethiopian perspective, if the frame is the number of published studies, plant ecology research is tremendously growing up. Even though evaluating the quality of research needs a different pathway, the quality of plant ecological research in Ethiopia is not growing up. The criteria here are the research theme composition, sophisticated technology and methods employed, and depth of the study. This is excluding the land use/land cover change domain that focused on the change detection only. Plant community, more or less, descriptive studies, dominated the plant ecology research thematic. This topic has been the center of plant ecological research in the 1960–1970s in Europe and Northern America. Hence, although it seems gloomy to bring this thought to the academia, plant ecology in Ethiopia is still in the 1970s.

11.3 Conclusions and Future Directions

This study systematically synthesized trends of plant ecology research in Ethiopia over the last 50 years. Descriptive studies on plant community ecology dominated the plant ecology research thematic most of them distributed in the DAF and MAF, while the Afroalpine and Ericaceous Belt, woodland (ACB and CTW), desert and semi-desert scrubland (DSS) vegetation types, and grasslands got little attention. I can say this is an "*Everything is in and from the forest*" syndrome. Open biomes (Bond 2019; Silveira et al. 2021) such as grasslands, shrublands, and woodlands deserve to be recognized in ecological restoration. Hence, the following future directions are suggested to improve the forthcoming plant ecology research in Ethiopia.

1. Future plant ecology research should gear to contemporary ecological researches. These include the application of remote sensing in vegetation ecology, climate change and vegetation ecology, plant functional ecology, vegetation temporal dynamics, and experimental approaches.
2. Experimental plant ecological studies are almost non-extant in the documents reviewed. Hence, future research should look into this perspective.
3. Establishing a committee that comprised plant taxonomists, plant ecologists, geologists, geographers, and GIS experts is recommended to investigate and map the plant community types at national level. This is crucial to allocate conservation resources objectively. Otherwise, fragmented studies of plant community types of a particular sites made conservation efforts challenging.
4. Recommendations from any research study, if available, should be based on the empirical information in order to make policy and decision justifiable.
5. Funders such as government agencies, NGOs, others including international ones are advised to provide resources to cover important topics such as invasive species ecology, application of GIS and remote sensing in vegetation ecology, and community interactions. Supervisors should also take the lion's share in helping graduate students identify contemporary ecological research areas such as pollination ecology.
6. The Ethiopian Ministry of Science and Higher Education (MoSHE) should establish a database both for completed and ongoing research projects, i.e., project registration in addition to what exists at various universities. This would help to avoid project redundancy.

References

Abatena H (1987) The potential contribution of indigenous self help and mutual aid organizations to the socio-economic development of Ethiopia. J Rural Coop 15:89–106

Arafaine Z, Asefa A (2019) Dynamics of land use and land cover in the Kafta-Sheraro national park, NW Ethiopia: patterns, causes and management implications. Momona Ethiop J Sci 11:239–257. https://doi.org/10.4314/mejs.v11i2.5

Asselin H, Gagnon D (2015) Trends in ecological research: reflecting on 21 years of Écoscience. Écoscience 22:1–5. https://doi.org/10.1080/11956860.2015.1120511

Barlow J, Stephens PA, Bode M et al (2018) On the extinction of the single- authored paper: the causes and consequences of increasingly collaborative applied ecological research. J Appl Ecol 55:1–4. https://doi.org/10.1111/1365-2664.13040

Beals EW (1969) Vegetational change along altitudinal gradients: studies in Ethiopia show that discreteness of zonation varies with steepness of slope. Sci New Ser 165:981–985

Bond WJ (2019) Open ecosystems: ecology and evolution beyond the forest edge. Oxford University Press, UK

Burivalova Z, Miteva D, Salafsky N et al (2019) Evidence types and trends in tropical forest conservation literature. Trends Ecol Evol 34:669–679. https://doi.org/10.1016/j.tree.2019.03.002

Carmel Y, Kent R, Bar-massada A et al (2013) Trends in ecological research during the last three decades—a systematic review. PLoS Med 8:e59813. https://doi.org/10.1371/journal.pone.0059813

Ferraro PJ, Pattanayak S (2006) Money for nothing? A call for empirical evaluation of biodiversity conservation investments. PLoS Biol 4:0482–0488

Friis I, Demissew S, van Breugel P (2010) Atlas of the potential vegetation of Ethiopia. The Royal Danish Academy of Sciences and Letters, Denmark

Game E (2018) Cross-discipline evidence principles for sustainability policy. Nat Sustain 1:452–454

Gebrehiwot K, Woldu Z, Fekadu M et al (2020) Classification and ordination of plant communities in Abune Yosef mountain range, northern Ethiopia. Acta Ecol Sin 40:398–411. https://doi.org/10.1016/j.chnaes.2019.12.001

Geeraert L, Aerts R, Berecha G et al (2020) Effects of landscape composition on bee communities and coffee pollination in Coffea arabica production forests in southwestern Ethiopia. Agric Ecosyst Environ 288:106706

Grace J (2019) Has ecology grown up? Plant Ecol Divers 12:387–405. https://doi.org/10.1080/17550874.2019.1638464

Hagen JB (2010) History of plant ecology. Encycl Life Sci 1–5

Harper JL (1967) A Darwinian approach to plant ecology. J Ecol 55:247–270

Harper JL (1977) Population biology of plants. Academic Press, London

Henderson PA (2012) Ecological methods. Encycl Life Sci 1–6

Hoffman M, Koenig K, Bunting G, et al (2016) Biodiversity hotspots (version 2016.1). https://doi.org/10.5281/ZENODO.3261807

Hundera K, Bekele T, Kelbessa E (2007) Floristics and phytogeographic synopsis of a dry afromontane confirous forest in the Bale mountain (Ethiopia): implications to biodiversity conservation. Sinet Ethiop J Sci 30:1–12

Hundera K, Deboch B (2008) Woody species composition and structure of the Gurra Farda Forest, Snnpr, South Wastern Ethiopia. Ethiop J Educ Sci 3

Kim JY, Joo GJ, Do Y (2018) Through 100 years of ecological society of America publications: development of ecological research topics and scientific collaborations. Ecosphere 9:e02109. https://doi.org/10.1002/ecs2.2109

Knight TM, Steets JA, Vamosi JC et al (2005) Pollen limitation of plant reproduction: ecological and evolutionary causes and consequences. Annu Rev Ecol Evol Syst 36:467–497

McCallen E, Knott J, Nunez-mir G et al (2019) Trends in ecology: shifts in ecological research themes over the past four decades. Front Ecol Environhttps://doi.org/10.1002/fee.1993

Miteva DA (2012) Evaluation of biodiversity policy instruments: what works and what doesn't? Oxford Rev Econ Policy 28:69–92

Mittermeier RA, Gil PR, Hoffmann M et al (2004) Hotspots revisited. CEMEX, UK

Mueller-Dombois D, Ellenberg H (1974) Aims and methods of vegetation ecology. Wiley, Toronto, Canada

Nobis M, Wohlgemuth T (2004) Trend words in ecological core journals o v er the last 25 years (1978–2002). Oik 106:411–421

Nune S, Soromessa T, Teketay D (2016) Land use and land cover change in the Bale Mountain eco-region of Ethiopia during 1985 to 2015. Land 5:41. https://doi.org/10.3390/land5040041

O'Connor RJ (2000) Why ecology lags behind biology. Science 14:35

Pellissier L, Alvarez N, Guisan A (2012) Pollinators as drivers of plant distribution and assemblage into communities. In: Patiny S (ed) Evolution of plant–pollinator relationships. Cambridge University Press, Cambridge, UK.

Peters RH (1991) A CRITIQUE FOR ECOLOGY. Cambridge University Press, Cambridge, UK

Pichi-Sermolli RE (1957) Una carta geobotanica dell' Africa orientale (Eritrea, Etiopia, Somalia). Webbia 13:15–132

Réale D, Khelfaoui M, Olivier P, Yves M (2020) Mapping the dynamics of research networks in ecology and evolution using co-citation analysis (1975–2014). Scientometrics. https://doi.org/10.1007/s11192-019-03340-4

Samnegård U, Hambäck PA, Lemessa D et al (2016) A heterogeneous landscape does not guarantee high crop pollination. Proc R Soc B Biol Sci 283:20161472

Samnegård U, Hambäck PA, Nemomissa S, Hylander K (2014) Dominance of the semi-wild honeybee as coffee pollinator across a gradient of shade-tree structure in Ethiopia. J Trop Ecol 30:401–408

Shiferaw H (2019) Performances of machine learning algorithms for mapping fractional cover of an invasive plant species in a dryland ecosystem. 2562–2574. https://doi.org/10.1002/ece3.4919

Shiferaw H, Bewket W, Alamirew T et al (2019) Implications of land use/land cover dynamics and Prosopis invasion on ecosystem service values in Afar Region, Ethiopia. Sci Total Environ Science 67:354–436. https://doi.org/10.1016/j.scitotenv.2019.04.220

Silveira FAO, Ordóñez-Parra CA, Moura LC, et al (2021) Biome awareness disparity is BAD for tropical ecosystem conservation and restoration. J Appl Ecol 00:1–9.https://doi.org/10.1111/1365-2664.14060

Teketay D, Bekele T (1995) Floristic composition of Wof-Washa natural forest, Central Ethiopia: implications for the conservation of biodiversity. Feddes Repert 106:127–147

UNEP-WCMC (2017) World database on protected areas user manual 1.5. Technical report. Cambridge, UK

van Breugel P, Kindt R, Lillesø JP et al (2015) Potential natural vegetation map of Eastern Africa (Burundi, Ethiopia, Kenya, Malawi, Rwanda, Tanzania, Uganda and Zambia). Version 2.0. Forest and Landscape Denmark, World Agroforestry Centre

Wassie A, Sterck FJ, Teketay D, Bongers F (2009) Tree regeneration in Church forests of Ethiopia: effects of microsites and management. Biotropica 41:110–119. https://doi.org/10.1111/j.1744-7429.2008.00449.x

Wassie SB (2020) Natural resource degradation tendencies in Ethiopia: a review. Environ Syst Res 9:33. https://doi.org/10.1186/s40068-020-00194-1

Woldu Z (1999) Forest in the vegetation types of Ethiopia and their status in the geographical context. In: Edwards S, Demissie A, Bekelle Y, Haase G (eds) forest genetic resources-conservation: principles strategies and action. Addis Ababa, Ethiopia

Woldu Z, Saleem MAM (2000) Grazing induced biodiversity in the highland ecozone of East Africa. Agric Ecosyst Environ 79:43–52